Praise for *Biocosm*

"A fascinating book, an eloquent and poetic synthesis of current ideas on the emergence of our biofriendly cosmos and its destiny. James Gardner's 'Selfish Biocosm' hypothesis envisions a novel perspective on humankind's role in the universe."

—**Sir Martin Rees,** Royal Society professor at Cambridge University; U.K. Astronomer Royal; author of *Just Six Numbers*, *Our Cosmic Habitat*, and *Our Final Hour*; and winner of the 2001 Gruber Prize in Cosmology

"James Gardner tackles the biggest of the Big Questions head on: Why is the universe bio-friendly? This stunning fact cannot be shrugged aside as an incidental quirk of nature, but deserves a deep and satisfying explanation. Gardner skillfully interweaves some of the most provocative ideas at the forefront of science to outline a possible explanation—and how extraordinary his explanation turns out to be!"

—**Paul Davies,** Professor of Natural Philosophy at the Australian Center for Astrobiology, author of *How to Build a Time Machine*, and winner of the Templeton Prize

"A magnificent one-stop account of the history of life. From the beginning of the universe to its end, Jim Gardner tells the entire story in a spell-binding account of how we got here and where we're going."

—**John Casti,** mathematician; author of *The Cambridge Quintet* and *The One, True, Platonic Heaven*; member of the External Faculty of the Santa Fe Institute; and former Executive Editor of *Complexity*, the journal of the Santa Fe Institute

"If there is no God—no outside transcendent being who designed and created the cosmos and life—from whence did it all come and how are we to find meaning in an apparently meaningless universe? The answer is derived from science, specifically the new sciences of chaos and complexity theory, that attempt to formulate natural explanations for these apparent supernatural phenomena. In this creative consilience of cosmology, evolutionary biology, and complexity theory, James Gardner courageously speculates about how it all could have come about and what it could possibly all mean using only the tools of science. *Biocosm* is breathtaking in its scope and its subject—the cosmos and everything in it—is far grander than the anthropocentric proscenium on which theistic world views play themselves out."

—**Michael Shermer,** Publisher of *Skeptic* magazine, monthly columnist for *Scientific American*, and author of *Why People Believe Weird Things*

BIOCOSM

BIOCOSM

THE NEW SCIENTIFIC THEORY OF EVOLUTION: INTELLIGENT LIFE IS THE ARCHITECT OF THE UNIVERSE

JAMES N. GARDNER

INNER
OCEAN

Inner Ocean Publishing, Inc.
P.O. Box 1239
Makawao, Maui, HI 96768-1239

Cover design: Bill Greaves
Cover illustration: Peter Crowther, Debut Art
Interior page design: Bill Greaves
Interior page typography: Madonna Gauding
Copy editor: Barbara Doern Drew

Publisher Cataloging-in-Publication Data

Gardner, James N.

Biocosm : the new scientific theory of evolution : intelligent life is the architect of the universe / James N. Gardner ; foreword by Seth Shostak. —1st ed.—Makawao, HI : Inner Ocean, 2003.

p. ; cm.

Includes bibliographical references and index.
ISBN 1-930722-26-5
ISBN 1-930722-22-2 (pbk.)
1. Evolution. 2. Cosmology. 3. Creation. I. Title.

B818.G37 2003
116–dc21 0309 CIP

Printed in Canada by Transcontinental
Distributed by Publishers Group West

9 8 7 6 5 4 3 2 1

This book is dedicated to the millions of men and women from every nation and culture who selflessly dedicate countless hours of unpaid time and other personal resources to the advancement of science. Whether as teaching assistants in elementary classrooms, docents in natural history museums, backyard stargazers, bird-watchers heedless of an early morning's predawn chill, uncompensated laborers on hot and dusty archaeological digs, or the more than four million SETI enthusiasts who happily link their home computers together to scan the random radio signals of the cosmos in hopes of finding the voice of ET, these volunteers are the unsung heroes and heroines of our scientific age. To them belongs the proud appellation derived from amator, *the Latin term for lover: "amateur."*

Nothing is too wonderful to be true
if it be consistent with the laws of nature.
—Michael Faraday (1791–1867)

Faraday was a British physicist and chemist who discovered electrolysis and invented both the transformer and the electric generator. The quoted statement was his response to ridicule expressed by his scientific contemporaries at his claim that he could generate an electric current simply by moving a magnet in a coil of wire.

Contents

Sidebars

Acknowledgments

Many helped me, in one way or another, in conceiving and writing this book. So a tip of the hat (in no particular order) to the following:

John Casti and Harold Morowitz, who took a chance on an unknown complexity theorist and published his papers in the prestigious Santa Fe Institute journal, *Complexity;*

Natasha Kern, literary agent extraordinaire, who never lost faith in me or this book;

Roger Jellinek, an exceptional editor whose vision and perceptiveness I came to admire deeply;

All the other great people at Inner Ocean Publishing, a small press bound for greatness—and especially my incredibly meticulous copyeditor, Barbara Doern Drew;

The distinguished scientists and academicians (including Freeman Dyson, John Barrow, Stuart Kauffman, Lee Smolin, Louis Crane, Dan Dennett, Andrei Linde, Harold Morowitz, and Sir Martin Rees) who read the manuscript or the essays on which it is based and offered their insights and sometimes brutal critiques (any errors that remain are solely my responsibility);

Jill Tarter, Tom Pierson, Seth Shostak, and their colleagues at the SETI Institute who graciously invited me to present my radical new theory to the SETI II Session of the International Astronautical Congress in Rio de Janeiro in 2000, and the John Templeton Foundation, which provided financial support for that presentation;

My brilliant and lovely wife, Lynda—the love of my life—who is an unfailing source of comfort, cheer, and inspiration;

And, finally, to the great, beautiful, and mysterious cosmos itself, which stands before us like a vast open book, written in a language we are just now beginning to understand.

Foreword

Is intelligent life merely a bit player in the enormous pageant of the cosmos? Or is it destined to become something vastly more important: the architect of the universe, and of other universes to come?

That is the striking question addressed in *Biocosm*, by James Gardner. While the ancient Greeks could summon enough of their dreaded hubris to imagine that humankind was the center of the cosmos, centuries of science have encouraged us to have a more modest view. Ever since Galileo, astronomers have somberly charted a universe that is stupefyingly large, bitterly cold, and implacably hostile. As every schoolchild knows, we occupy a small planet around a common sort of star, itself just one of a hundred billion suns in a rather ordinary galaxy. Our cosmic situation is insignificant, and life, particularly intelligent life, might be only an accident of circumstance on one tiny, watery world. Indeed, we have yet to discover whether even a single other celestial body has managed to spawn the simplest biology.

In other words, we might be tempted to judge from current evidence that life is no more than an occasional consequence of nature's laws, a chance product of the chemistry that those laws allow. Intelligence, which has emerged only recently on our planet, might be even rarer. In this view, the appearance of life seems if not miraculous, then at least highly unlikely.

But there is something wrong with this picture. As our understanding of cosmology has deepened, we have been confronted with a disarming fact: it seems that the physical laws and constants of the universe have been finely tuned for life. For example, the energy states of atoms are such as to allow the easy formation of carbon in the searing interiors of the stars, and to prevent this element from being quickly transmuted into yet heavier atoms. But there seems to be no compelling reason why these energy levels could not have been otherwise, resulting in a universe in which carbon—the key building block of complex molecules and therefore of life —was hard to find. The particulars of the Big Bang were also fortuitous. Had this initial event occurred with less force, the universe would have long ago collapsed on itself. With more force, it would have expanded too quickly to allow the formation of galaxies, stars, planets—and us. Again, it is hard to explain why, like Goldilocks's porridge, the Big Bang should have detonated with a force that was "just right." In the words of physicist Freeman Dyson, such apparent coincidences make it seem as if "the universe knew we were coming."

This is not a trivial observation, nor have there been broadly satisfactory explanations offered for why the cosmos is so hospitable, despite the best efforts of eminent scientists. One straightforward approach is to invoke a deity who deliberately engineered our universe in such a way that life could arise. But such an explanation seems difficult or impossible to prove. Perhaps the "fine tuning" that we perceive is only a temporary condition brought on by our incomplete knowledge of physics. With another century of research, this argument goes, we might understand that the universe had no choice in its construction: the energy levels that result in abundant carbon could not have had other values. Still another tack is to simply note that if the universe had not been tuned for life, we would not be here to write book forewords that ask *why*. Needless to say, this may be too tautologically trivial to appeal to many.

James Gardner carefully reviews all the best ideas on how to understand the cosmos's apparent biological imperative and then puts forth a new, and strikingly dramatic, suggestion of his own, one that makes use of the exciting field of complexity science. He is well qualified to do this, with training in theoretical biology and philosophy, and an impressive trail of published, scholarly work in complexity theory. His arguments are lucid, and his prose is elegant and engaging. But what will most strike the reader of this book is the fact that Gardner is not going after small fish. The subject he is wrestling with is as large, as important, as they come: What is the purpose of our universe and the life it has spawned? He tells us how the fact that the universe was "made for life" can be ultimately understood by science and need not forever be the domain of theology or metaphysics.

As an astronomer engaged in the search for extraterrestrial intelligence (SETI), I had occasion to meet Gardner at an international conference several years ago. I watched as his proposals about a "selfish biocosm" caught the attention of my colleagues. I think their interest was principally piqued by the freshness of his thought and the broad sweep of his ideas. But there is more: one of the important tests of Gardner's hypothesis is the SETI enterprise. If we should soon find evidence of alien intelligence, that would be an important datum in Gardner's new way of assessing the universe.

Ever since Newton, scientists have tried to understand existence by discovering its underlying rules. The result has been a massive edifice of natural law, and biology has been seen as a consequence of the universe's construction, rather than an instigator. Only on Earth's surface, where life has molded the seas, the continents, and even the atmosphere, is biology thought to have had an important role in shaping physical conditions—the so-called Gaia hypothesis. But Gardner has taken Gaia to its furthest conceivable magnitude: extending the role and influence of life to the stars and beyond.

There is little doubt that his ideas will change yours.

Seth Shostak
Mountain View, California

Introduction

This book presents a new theory about the role of life and mind in shaping the origin and ultimate fate of the universe. In addition, it reflects on how that new theory might eventually influence religion, ethics, and our self-image as a species.

In important respects, my book is a riff on Charles Darwin's masterwork, *The Origin of Species.* Following Darwin's lead, I have endeavored to use the insights proffered by a wide range of gifted contemporary theorists—cosmologists, evolutionary biologists, computer scientists, and complexologists—to construct the foundation for a novel and somewhat startling synthesis. The essence of that synthesis is that life, mind, and the fate of the cosmos are intimately and indissolubly linked in a very special way. To echo the insightful phrase of Princeton astrophysicist Freeman Dyson, it is my contention that "mind and intelligence are woven into the fabric of our universe in a way that altogether surpasses our comprehension."[1]

The fundamental credo of science is that physical mysteries that presently elude human understanding will someday, if only in the far distant future, succumb to new explanatory paradigms that are capable of being either validated or discredited through falsifiable predictions. (Falsifiability of claims, which is scientific shorthand for the empirical testability of new hypotheses and their implications, is the hallmark of genuine science, sharply demarcating it from other arenas of human thought and experience like religion, mysticism, and metaphysics.) The basic claim of this book is that the oddly life-friendly character of the fundamental physical laws and constants that prevail in our universe can be explained as the predictable outcome of natural processes—

specifically the evolution of life and intelligence over tens of billions of years.

The explanation that I shall put forward to elucidate the linkage between biological evolution and the ultimate fate of the cosmos—a new theory called the "Selfish Biocosm" hypothesis—has been developed in papers and essays published in peer-reviewed scientific journals like *Complexity* (the journal of the Santa Fe Institute, the leading center for the study of the new sciences of complexity), *Acta Astronautica* (the journal of the International Academy of Astronautics), and the *Journal of the British Interplanetary Society.* These papers provide the foundation for a scientifically plausible version of the "strong anthropic principle"—the notion that the physical laws and constants of nature are cunningly structured in such a way as to coax the emergence of life and intelligence from inanimate matter.

The book is divided into six parts. The first part reviews the profound mysteries of an anthropic—or life-friendly—universe. Beginning with ancient Greek philosophy, continuing on through Renaissance thought, and concluding with contemporary speculations by a leading complexity theorist about a mysterious antichaotic force in nature, this section provides the foundation for theoretical speculations about possible reasons why the universe is life-friendly.

The second part of the book plunges deeper into the anthropic mystery and probes some of the novel ideas that contemporary scientists have advanced by way of explanation. These include the conjecture by a leading cosmologist that black holes are gateways to new universes.

The third part makes a risky foray into the dangerous territory that is the situs of the contemporary cultural war between ultraevolutionists and modern creationists, who call themselves "intelligent design theorists." In a proposed harmonization of these conflicting viewpoints, I suggest that the appearance of cosmic design could conceivably emerge from the operation of evolutionary forces operating at unexpectedly large scales.

The fourth part of the book puts forward my new Selfish Biocosm hypothesis: that the anthropic qualities that our universe exhibits can be explained as incidental consequences of an enormously lengthy cosmic replication cycle in which a cosmologically extended biosphere provides the means by which our cosmos duplicates itself and propagates one or more "baby universes." The hypothesis suggests that the cosmos is "selfish" in the same metaphorical sense that evolutionary theorist and ultra-Darwinist Richard Dawkins proposed that genes are "selfish."[2] Under my theory, the cosmos is "selfishly" focused upon the overarching objective of achieving its own replication. To use the terminology favored by economists, self-reproduction is the hypothesized "utility function" of the universe.

An implication of the Selfish Biocosm hypothesis is that the emergence of life and ever more accomplished forms of intelligence is inextricably linked to the physical birth, evolution, and reproduction of the cosmos. This section also provides a set of falsifiable implications by means of which the new hypothesis may be tested.

The fifth part of the book enters a more speculative realm by considering methods by which a sufficiently evolved form of intelligence might replicate the life-friendly physical laws and constants that prevail in our universe. In addition, it advances the idea that if the space-time continuum (i.e., our cosmos in its entirety) constitutes a closed loop linking one gateway of time (the Big Bang) to another (the Big Crunch), then our anthropic universe could conceivably, in the words of Princeton astrophysicist J. Richard Gott III, be its own mother.[3]

The sixth and final section ponders the possible implications of the Selfish Biocosm hypothesis for fundamental evolutionary theory and for our self-image as a species. It also takes a brief look at possible religious and ethical implications of the hypothesis.

A major caveat is in order before we begin. This book is intentionally and forthrightly speculative. Following the example of Darwin, I have attempted to crudely frame a revolutionary explanatory paradigm well before all of the required building materials and construction tools are at hand. Darwin had not the slightest clue, for instance, that DNA is the molecular device used by all life-forms to accomplish the feat of what he called "inheritance." Indeed, as cell biologist Kenneth R. Miller noted in *Finding Darwin's God*, "Charles Darwin worked in almost total ignorance of the fields we now call genetics, cell biology, molecular biology, and biochemistry."[4] Nonetheless, Darwin managed to put forward a plausible theoretical framework that succeeded magnificently despite the fact that it was utterly dependent on hypothesized but completely unknown mechanisms of genetic transmission.

As Darwin's example shows, plausible and deliberate speculation plays an essential role in the advancement of science. Speculation is the means by which new paradigms are initially constructed, to be either abandoned later as wrong-headed detours or vindicated as the seeds of scientific revolution.

Scientific speculation plays another equally important role, which is to shine the harsh light of skepticism on accepted verities. As the brilliant and controversial Cornell physicist Thomas Gold put it,

> New ideas in science are not right just because they are new. Nor are old ideas wrong just because they are old. A critical attitude is clearly required of every seeker of truth. But one must be *equally* critical of both the old ideas as of the new. Whenever the established ideas are accepted uncritically and conflicting new evidence is brushed aside or not even reported because it does not fit, that particular science is in deep trouble.[5]

Science is an inherently conservative discipline, and iconoclastic ideas like those entertained by Gold (the existence of a deep hot biosphere far beneath the planet's surface as well as the nonbiological origin of natural gas and oil) are legitimately relegated to what *Skeptic* magazine publisher Michael Shermer calls the borderlands of science.[6] But what must never be forgotten is that these dimly illuminated borderlands have frequently proven to be the breeding ground of revolutionary ideas.

Scientific revolutions differ profoundly in character from the normal practice of scientific investigation. Scientific historian Thomas Kuhn observed in his classic *The Structure of Scientific Revolutions* that normal science consists of puzzle solving within the framework provided by prevailing scientific paradigms (like Newtonian mechanics or Darwinian theory), which are themselves the fruit of earlier revolutions. Revolutionary science, by contrast, is a hazardous but utterly exhilarating process of creative destruction—the erection of fundamental new paradigms to supplant or supplement a foundational structure that has become hopelessly flawed.[7] As science popularizer James Gleick put it in *Chaos: Making a New Science*,

> Then there are the revolutions. A new science arises out of one that has reached a dead end. Often a revolution has an interdisciplinary character—its central discoveries often come from people straying outside the normal bounds of their specialties. The problems that obsess these theorists are not recognized as legitimate lines of inquiry. Thesis proposals are turned down or articles are refused publication. The theorists themselves are not sure whether they would recognize an answer if they saw one. They accept risk to their careers. A few freethinkers working alone, unable to explain where they are heading, afraid even to tell their colleagues what they are doing—that romantic image lies at the heart of Kuhn's scheme, and it has occurred in real life, time and time again.[8]

The borderlands of science, in short, are the natural habitats of scientific revolutionaries—those free-spirited souls who cheerfully risk professional ridicule in return for the sublime privilege of attempting to pull one more veil from nature's deeply shrouded visage.

For me, the pathway to the particular scientific borderland that is the subject of this book has meandered through the novel intellectual landscape illuminated by the new sciences of complexity. These sciences, which explore phenomena like "emergence" (the generation of complicated phenomena such as consciousness from the interaction of relatively simple components like individual nerve cells), self-organization, and the operation of complex adaptive systems (like sets of coevolving species comprising a biosphere), have generated not only scholarly excitement but a rapidly rising level of popular interest. The great appeal of these sciences is their inherently holistic quality, so different from the reductionist approach favored by practitioners of so-called hard physical sciences like physics and chemistry. These traditional sciences tend to foster a "silo" mentality that frowns on cross-disciplinary thinking. By contrast, scientists studying complexity deliberately seek out the recurrence of similar patterns of evolutionary development and emergence in a wide range of seemingly disconnected phenomena, from embryology to cultural evolution and from theoretical chemistry to the origin of life.

The key experimental tool utilized by complexologists is not physical measurement but computer simulation; the "experiments" of complexity scientists generally take place in what mathematician John Casti calls "would-be worlds" that exist only in the memory and logic chips of a computer.[9] As Casti puts it, "With our newfound ability to create worlds for all occasions inside the computer, we can play myriad sorts of what-if games with genuine complex systems. No longer do we have to break the system into simpler subsystems or avoid experimentation completely because the experiments are too costly, too impractical, or just plain too dangerous."[10]

The holistic philosophy embodied in the sciences of complexity is uniquely suited to the mission of the intellectual voyage on which we shall presently embark: to seek out and delineate, as precisely and exhaustively as possible, a specific theory concerning the linkage and "consilience" (in biologist Edward O. Wilson's resonant phrase)[11] between the basic laws and constants governing the behavior of inanimate nature and the role of life and mind in the universe. As we shall see, the very fact that such consilience and linkage should exist is itself a profound ontological commentary.

Now, why am I—an attorney, a complexity theorist, and a science essayist—qualified to serve as your guide on this daunting journey to the outer limits of cosmological theory? In part because, as an attorney, I am trained to search for faint and elusive patterns of evidence that a layperson might overlook—including evidence that crosses traditional disciplinary lines demarcating the borders of disparate scientific fields.

I first began probing the mysteries of complexity theory in a scholarly paper that proposed an interpretation of the behavior of subnational geopolitical regions (like Flanders in Belgium and Catalonia in Spain) as the operation of complex adaptive systems.[12] After this essay was published in *Complexity*, I turned my attention to another set of complex phenomena: the probable future coevolution of "memes" (hypothetical units of cultural transmission) and genes in the context of the rapidly emerging technological capacity to engage in human germline genetic engineering. That essay—which is reproduced here in appendix 1—was likewise published in *Complexity*.[13]

With that foundation in place, I decided to use the approach of complexity theory to probe an odd feature of cosmology that has intrigued me ever since I began studying philosophy and theoretical biology as an undergraduate at Yale: the strangely life-friendly quality of the physical laws and constants that prevail in our universe. As a lawyer, I was goaded by the sense that the patterns of evidence seemed to be pointing in a direction that most mainstream scientists were unwilling to explore. As a student of philosophy and biology, I was convinced that issues of profound importance were being overlooked or deliberately shunned. And as a recent convert to the holistic philosophy represented by the sciences of complexity, I was becoming increasingly convinced that the pathway to genuine enlightenment about the import of an anthropic universe—a universe adapted to the needs of life just as thoroughly as life is adapted to the exigencies imposed by the universe—must surely pass through the strange and intriguing intellectual terrain revealed by these new sciences.

I explored that possibility in an essay published in *Complexity* entitled "The Selfish Biocosm: Complexity as Cosmology."[14] I was privileged to have as the chief reviewer for this paper an individual who is one of the most distinguished theoretical cosmologists in the world. And I was equally privileged to have the services of a courageous editor—John Casti—who was willing to take a chance on a relatively unknown theorist advancing a radically new hypothesis about the intimate relationship of life and intelligence to fundamental cosmic forces and laws. That essay, of which this book is an expanded and augmented version, was my first attempt to crudely map out what is, for me at least, a singularly exciting new borderland of science.

Like a medieval European map maker piecing together the borders of an imagined America from travelers' tales and the misty recollections of ancient mariners, my role (at least as I perceive it) is not to serve as an explorer or experimentalist but rather to sketch the larger features of a vision of cosmic reality profoundly at odds with traditional wisdom. In medieval times, the orthodox view was that the surface of Earth was flat. In the contemporary era, the prevailing scientific mindset is captured curtly and elegantly by Nobelist Steven Weinberg's pithy epigram that "the more the universe seems comprehensible, the more it also seems pointless."[15] It is my fervent hope that those who consider seriously the speculative exercise in intellectual cartography presented in this book will conclude that Weinberg's assertion may eventually prove to be as mistaken as the flat-Earth orthodoxy espoused with such strenuous but utterly misplaced confidence in a bygone age.

With that preface, I invite you to enter what I believe to be the least tamed and most challenging scientific borderland of all: current theorizing about the ultimate nature and destiny of the vast cosmos that envelops our tiny speck of Earth like an endless sea. Perhaps you will find in the speculative discourse that follows some useful nugget of fact or some momentary flash of insight that helps pierce, to at least a minuscule degree, the perplexing darkness that surrounds the outer ramparts of twenty-first-century cosmological science. If so, I will have succeeded in communicating a faint echo of the sense of wonder and awe at the abiding mysteries of nature so perfectly captured by Isaac Newton three hundred years ago: "I do not know what I may appear to the world, but to myself I seem to have been only like a boy playing on the seashore, and diverting myself in now and then finding a smoother pebble or a prettier shell than ordinary, whilst the great ocean of truth lay undiscovered before me."[16]

Part One

The Profound Mystery of an Anthropic Universe

As explained in the introduction, this book begins with three chapters that attempt to lay the groundwork necessary to begin thinking about the basic nature of the universe in a radically new way. At the heart of this new way of thinking is an attempt to take seriously what strikes many scientists as an irksome curiosity: the fact that the universe is, against spectacular odds, strangely life-friendly or "anthropic" (to use the term favored by cosmologists).

Part I reviews the thoughts of leading philosophers and scientists from antiquity to the present era on this mysterious and neglected topic.

BIOCOSM

Chapter One

Searching for ET in All the Wrong Places

The search for extraterrestrial intelligence (ET) is easily the most exotic of mainstream scientific pursuits. Inspired by the famous Drake equation—which posits that the incidence of scientifically literate ET civilizations in our galaxy can be predicted by a simple mathematical formula—the quest to identify a radio or optical signal that would qualify as a genuine alien artifact has captured the imagination of a motley crew of scientists, billionaires, novelists, and ordinary citizens ever since SETI pioneer Dr. Frank Drake first conceived it in 1961. (See sidebar, page 3.)

The most curious aspect of this unique endeavor is its notorious lack of even partial success. As physicist Enrico Fermi commented during a luncheon conversation in Los Alamos in 1950, noting the conspicuous absence of any indication of the existence of intelligent alien life, "Where are they?"[2] The overwhelming consensus of the scientific community is that we have not yet come across any credible evidence of extraterrestrial intelligence. The SETI Institute—the informal worldwide headquarters for the search for extraterrestrial intelligence—concedes that it has recorded no artificial signal that qualifies as hard evidence of ET's existence. UFO sightings are universally discredited by serious scientists. Those that remain unexplained are simply ignored—and for good reason. As the late astronomer Carl Sagan pointed out, extraordinary scientific hypotheses, like claims of alien visitation, impose extraordinary proof requirements.

But is it possible that we have been searching for signs of extraterrestrial intelligence in all the wrong places? Could the telltale evidence be right under our noses? To begin to answer this question, it is well to remember that in Sagan's memorable novel, *Contact*, there were actually

The Drake Equation

In 1961, Dr. Frank Drake, a pioneering researcher in the field of SETI (the Search for Extraterrestrial Intelligence), developed a formula that articulates the various factors that make the existence of technological civilizations beyond Earth more or less probable.[1] *The formula, which is obviously speculative in the extreme, is really nothing more than a systematic way of quantifying our ignorance, in the felicitous phrase of SETI researcher Jill Tarter.*

The Drake equation is as follows:

($N = R f_p n_e f_l f_i f_c L$)

The symbols mean the following:

- *N is the number of civilizations in the Milky Way galaxy capable of communicating across*

(continued on next page)

two independent indications of alien sentience. The first, upon which the plot is centered, was the radio signal intercepted by Jody Foster's movie character (modeled on the real-life Jill Tarter, head of the SETI Institute's Project Phoenix).

The second piece of evidence was far more cryptic and interesting. It was a kind of artist's signature—a recipe for drawing a perfect circle in base 11 arithmetic—inscribed many kilometers to the right of the decimal point in the infinite numerical expanse of pi.[3] (Base 11 arithmetic differs from commonly used base 10 arithmetic by employing a scale of eleven rather than a scale of ten for counting and scaling.) This too was a kind of signal—an unmistakable artifact of extraterrestrial intelligence—woven with consummate subtlety into the mathematical fabric of our reality.

Is it possible that this is the category of extraterrestrial (or, perhaps better phrased, *trans*terrestrial) intelligence for which we ought to be searching? Might it be that we have already stumbled across evidence of its existence, without ever realizing the import of our discovery?

This astonishing possibility was lyrically addressed by the late physicist Heinz Pagels in *The Dreams of Reason.*[4] In a cryptic and largely overlooked passage, he raised this portentous possibility:

> The SETI astronomers hope to hear intelligent signals. Other astronomers think that we are alone in the galaxy and that any such search is a waste of scarce resources. It is hard to prejudge the prospects of SETI's success—the project is an exciting long shot in any case. I'm not sure if looking for an alien intelligence in outer space is the only place to look. I think that there is One here already.[5]

Pagels was a wonderfully gifted scientist and science popularizer who met an untimely demise in 1988 in a mountain-climbing accident on Pyramid Peak in Aspen, Colorado. As president of the prestigious New York Academy of Sciences at the time of his death, Pagels was one of the few mainstream intellectuals willing to contemplate the deeper implications of the very existence of what he called a "cosmic code"—

interstellar distances via radio signals or other technological means.

- R *is the rate at which stars capable of sustaining life are born.*
- f_p *is the percentage of those stars that are accompanied by planetary systems.*
- n_e *is the number of planets in a planetary system that are terrestrial (i.e., small and rocky like Earth as opposed to giant and gaseous like Jupiter or Saturn).*
- f_l *is the percentage of such planets that harbor life.*
- f_i *is the percentage of the preceding subset that are home to intelligent life.*
- f_c *is the percentage of this still smaller fraction of planets that host life-forms capable of using radio or other technology (like lasers) to communicate over interstellar distance.*
- L *is the average lifetime of such advanced civilizations.*

Some of the factors are more susceptible to scientific estimation than others. For instance, the recent success of astronomers in locating hordes of so-called *exo-planets orbiting distant stars offers real hope that scientists will soon be able to quantify in a statistically meaningful way the f_p factors. Others, like L and f_c seem intractably elusive.*

the oddly coherent ensemble of scientific laws and physical constants that govern the behavior of everything from subatomic particles to giant structures comprising thousands of galaxies.

In his book, Pagels asked us to imagine a creator, the Demiurge, "much as most medieval Europeans did—a Being who made the universe as his personal creation."[6] While Pagels, like most scientists, rejected belief in a personal creator-God as "hopelessly inadequate and without evidence," he was nonetheless willing to "entertain the notion of the Demiurge for the sake of our argument."[7] As Pagels put it, "After all, you don't have to *believe* in the Demiurge to *imagine* that a Being made the stars, Earth, and moon, the planets, animals, and us—surely a highly intelligent Being."[8]

With this hypothesis in hand, Pagels proceeded to lead his readers into exotic intellectual terrain largely unexplored by traditional scientists and most theologians. He asked us to contemplate the possibility of a unique communication system capable of transmitting signals from a vastly superior extraterrestrial intelligence located far away in space and time to mere mortals on Earth and other habitable planets: "For my purposes, the Demiurge is a good model for the Alien Intelligence from whom we are currently getting intelligent signals. How does the communication system work? Let us first examine a simple example of the kind of communication system that I have in mind."[9]

Pagels's vision of this ingenious message transmission system was based on an archaeological analogy:

> Suppose a group of archaeologists discover the ruins of an ancient civilization buried underground and covered by jungle. At first they know almost nothing—just the fact that the civilization existed. But as they clear away the ground and jungle they find buildings, temples, and tombs. Slowly, over a period of decades, from the artifacts and pictorial inscriptions, they begin to piece together the history of an ancient people—they are interpreting the information inherent in the ruins. The ruins may thus be viewed as a message, an intelligible structure, and the ancient people who built the civilization as the "alien

Spinoza and Einstein

In an oft-quoted statement, Albert Einstein said, "I believe in Spinoza's God who reveals himself in the harmony of all that exists, but not in a God who concerns himself with the fate and actions of human beings."[15] Indeed for Einstein, as for the seventeenth-century religious philosopher Baruch Spinoza whom he respected deeply, the concept of deity and the idea of profound order in the universe were interchangeable. As the great scientist put it in 1936, "Everyone who is seriously involved in the pursuit of science becomes convinced that a spirit is manifest in the laws of the universe—a spirit vastly superior to that of man."[16]

(continued on next page)

intelligence" from whom the archaeologists are receiving the message. One does not usually think of artifacts from the human past as a communications link, but that is indeed one way they function.[10]

Pagels then asked us to consider this image—archaeologists deciphering the intelligible message inherent in an ancient relic—as a metaphor for scientific investigation:

> Imagine now, instead of the ruins of an ancient civilization as the intelligible structure, that the material universe, the "ruins" of the hot big bang, contains a kind of message. The universe, after all, like the ruins, has a definite structure, it is intelligently organized, and that organization can be studied by natural scientists. The universe itself may thus be viewed as the communications link between the Alien Intelligence—the Demiurge who created it—and us.[11]

And what kind of a message from an unknown ET was being conveyed by means of this unique communications link? Pagels suggested that the message consists of nothing less than the full suite of rules and constants that make up the laws of nature:

> What [scientists] find is that the architecture of the universe is indeed built according to invisible universal rules, what I call the cosmic code—the building code of the Demiurge. . . . Scientists in discovering this code are deciphering the Demiurge's hidden message, the tricks he used in creating the universe. No human mind could have arranged for any message so flawlessly coherent, so strangely imaginative, and sometimes downright bizarre. It must be the work of an Alien Intelligence![12]

The oddest feature of this cryptic message, as Pagels saw it, was that "as far as we can tell, the Demiurge has written himself out of the code—an alien message without evidence of an alien."[13] While Pagels thus believed—and others will obviously disagree—that "we can safely

Einstein's vision of deity was not anthropomorphic in the slightest. In particular, Einstein disdained the vision of a vengeful God bent on punishing disobedient or nonbelieving denizens of his creation: "I cannot imagine a God who rewards and punishes the objects of his creation, whose purposes are modeled after our own—a God, in short, who is but a reflection of human frailty."[17]

For Einstein, the path toward genuine religious enlightenment and the road toward scientific insight were one and the same. As he put it, "The further the spiritual evolution of mankind advances, the more certain it seems to me that the path to genuine religiosity does not lie through the fear of life, and the fear of death, and blind faith, but through striving after rational knowledge."[18] Those who seek this path are, by and large, serious scientists operating at the very frontiers of their disciplines. As Einstein famously wrote in The New York Times *in 1930, speaking of the "cosmic religious feeling" shared by many leading scientific researchers around the world, "A contemporary has said, not unjustly, that in this materialistic age of ours the serious scientific workers are the only profoundly religious people."[19]*

drop the traditional idea of a Demiurge, for there is no scientific evidence for a Creator of the natural world, no evidence for a will or purpose in nature that goes beyond the known laws of nature,"[14] he nonetheless concluded that extraordinarily intriguing questions remain about the content and import of the extraterrestrial message itself, namely:

- Are the laws and constants of nature that comprise the cosmic code specified uniquely by mathematics and thus subject to no conceivable variation? Or might these laws and constants actually vary from universe to universe within a vast cosmic population known as a multiverse?

- If variation is possible, is our particular universe special inasmuch as its physical laws, its dimensional geometry, and the values of its physical constants are oddly hospitable to carbon-based life? If so, what conceivable natural process could have rendered our universe so life-friendly?

- If there is no cosmic creator apart from the message, could the message itself conceivably *be* the creator? And a corollary question: Did the deistic philosopher Baruch Spinoza, who influenced Einstein so profoundly, get it right when he characterized the search for reason and order in the universe as tantamount to the search for a cosmic creator? (See sidebar, page 5.)

These troubling questions straddle the uneasy frontier demarcating the domains of religion and science. Indeed, they are closely related to issues raised by the so-called strong version of the cosmological anthropic principle, a concept that many prominent scientists deride as a mere theological notion. (The strong anthropic principle, which Pagels quietly embraced, predicts that the origin of life and intelligence in the universe will eventually be shown to be strongly favored by the laws of nature. The less controversial weak version of the anthropic principle,

NASA's Origins Program

It may be a slight exaggeration to characterize NASA's view as that the presence of liquid water on a planet equates to the appearance of life, but it is certainly true that the space agency's own description of its Origins program is exuberantly optimistic about the likelihood that life will dependably arise whenever physical conditions are propitious. Here is NASA's own breathless description of the rationale for its life-seeking mission:

Approximately 15 billion years ago in cosmic history, the first galaxies took shape from vast clouds of early chemical

(continued on next page)

which Pagels scornfully disdained, merely states in tautological fashion that since human observers inhabit this particular universe, it must perforce be life-friendly or it would not contain any observers resembling ourselves.)

As this book will argue, the anthropic, or life-friendly, qualities of our cosmos—a spectacularly unlikely congeries of physical laws and constants that seem altogether too perfectly suited to the emergence and evolution of carbon-based life and intelligence to be the product of any conceivable random process—constitute both a profound mystery and a subtle set of clues pointing toward a possible solution. These clues indicate that the impression of design in nature is no mere illusion and that the predisposition of the cosmos to breathe life and intelligence into inanimate matter is deeply embedded in the organizing principles of nature. As Pagels put it, "My own view is that although we do not yet know the fundamental laws, when and if we find them the possibility of life in a universe governed by those laws will be written into them. The existence of life in the universe is not a selective principle acting upon the laws of nature; rather it is a consequence of them."[20]

When Heinz Pagels perished in the mountains of Colorado, modern science lost one of its great visionaries. But his spirit of intellectual adventure lingers on, and it provokes us to ask this simple question: Can we ever hope to employ the tools of science to probe the deep mysteries to which Pagels alluded, or will they be forever confined to the realm of metaphysics and religious faith?

It is the hypothesis of this book that science will eventually be capable of tackling these and many related issues that have long been viewed as religious in nature. I believe that scientists, like diligent cryptographers, will someday be able to decipher completely the commodious and utterly mysterious message from an unknown ET that Heinz Pagels described as the cosmic code.[21]

Why do I believe this? Mainly because of the historical success of the scientific research endeavor. The history of the ascendance of science is the chronicle of an incredible journey: an epic migration by the inquisitive human spirit from a dim and terrifying landscape of ignorance

elements. In the furnace of stars, life-sustaining chemicals such as carbon and oxygen came into being. Then, in awesome blasts from dying stars, life's chemicals blew out into space, only to condense anew into stars like our sun and planets like Earth. Through the mixing of these vital chemicals and energy, the living Universe blossomed from the earliest self-replicating organisms and the profusion of life on our planet. Seeing similar chemical conditions wherever we look in the cosmos, the hope of finding life somewhere else inevitably arises.[24]

NASA's Origins program actually encompasses four distinct science goals:

- *Goal 1: To understand how galaxies formed in the early universe.*
- *Goal 2: To understand how stars and planetary systems form and evolve.*
- *Goal 3: To determine whether habitable or life-bearing planets exist around nearby stars.*

- *Goal 4: To understand how life forms and evolves.*

The overriding goal of this audacious research initiative is to answer two enduring questions:

- *Where do we come from?*
- *Are we alone in the universe?*

Answering the first question means, in NASA's words, "understanding how the great chain of events unleashed after the Big Bang culminated in us and in everything we observe today. It is the story of our cosmic roots, told in terms of all that precedes us: the origin and development of galaxies, stars, planets, and the chemical conditions necessary to support life."

Answering the second question depends, according to NASA, on the success of "our search for life-sustaining planets and on our understanding of [life's] glorious diversity here on Earth."[25]

and superstition—a dark territory that Carl Sagan called the demon-haunted world—to a well-lit dominion where reason and lucidity hold sway. It is an awe-inspiring saga of mysteries solved and greater mysteries unveiled.

We now stand poised on the brink of a dramatic voyage of discovery of breathtaking import. It will be a journey like no other, for it promises to open up radically new vistas of insight into ancient mysteries. A scientifically rigorous formulation of the strong anthropic principle, I submit, may well provide the key intellectual scaffolding that will allow us to successfully attack these profound questions.

This book will tell the extraordinary story of a looming scientific revolution in cosmology, evolutionary theory, and the sciences of complexity. It will also present a testable cosmological hypothesis with truly astonishing implications.

The essence of what I am calling the "Selfish Biocosm" hypothesis is that the universe we are privileged to inhabit is literally in the process of transforming itself from inanimate to animate matter. As physical chemist and philosopher Michael Polanyi put it, "This universe is still dead, but it already has the capacity of coming to life."[22] Under this theory, the emergence of life and intelligence are not meaningless accidents in a hostile, largely lifeless cosmos but exist at the very heart of the vast machinery of creation, cosmological evolution, and cosmic replication.

Remarking on the paradigm-shattering power of this staggering possibility, the award-winning science writer Paul Davies said this in *The Fifth Miracle:*

> In claiming that water means life, NASA scientists are . . . making—tacitly—a huge and profound assumption about the nature of nature. [See sidebar, page 7.] They are saying, in effect, that the laws of the universe are cunningly contrived to coax life into being against the raw odds; that the mathematical principles of physics, in their elegant simplicity, somehow know in advance about life and its vast complexity. If life follows from [primordial] soup with causal dependability, the laws of nature encode a hidden subtext, a cosmic imperative, which

> tells them: "Make life!" And, through life, its by-products: mind, knowledge, understanding. It means that the laws of the universe have engineered their own comprehension. This is a breathtaking vision of nature, magnificent and uplifting in its majestic sweep. I hope it is correct. It would be wonderful if it were correct. But if it is, it represents a shift in the scientific world-view as profound as that initiated by Copernicus and Darwin put together.[23]

The evidence is mounting that the laws of nature—Pagels's cosmic code—do indeed encode a hidden subtext that scientists may actually be capable of deciphering. Respected experts like astrophysicists John Barrow and Frank Tipler, physicists Freeman Dyson and John Wheeler, Nobel Prize–winning biologist Christian de Duve, and British Astronomer Royal Sir Martin Rees as well as many other scientists have begun to soberly assess the astonishing array of "just so" coincidences inherent in the physical characteristics of our universe—qualities that render the cosmos spookily hospitable to carbon-based life. The scientists' incredible conclusion: the statistical improbability of a universe possessing all the life-friendly characteristics exhibited by our cosmos is simply too great to exclude the possibility of a nonrandom origin.

No one can yet say whether the emerging paradigm will permit humanity to finally begin to answer the portentous question posed by John Wheeler in 1989 in a celebrated Santa Fe Institute lecture entitled "It From Bit." He said, "A single question animates this report: Can we ever expect to understand existence? Clues we have, and work to do, to make headway on that issue. Surely someday, we can believe, we will grasp the central idea of it all as so simple, so beautiful, so compelling that we will all say to each other, 'Oh, how could it have been otherwise? How could we all have been so blind so long?'"[26] What we can reasonably anticipate, however, is that scientists will soon be prepared to launch a great quest, using the latest tools of cosmology and complexity theory to decipher important missing chapters of the cosmic code and seek out answers to such profound questions as the following that have gripped the minds of philosophers and prophets for thousands of years:

Point Omega

The concept of Point Omega (or the Omega Point) is associated with the religious philosopher Teilhard de Chardin and, in the early twenty-first century, with the work of cosmologists John Barrow and Frank Tipler. For Teilhard, the Omega Point was the predicted end point of terrestrial evolution—an epoch in the distant future when humanity will have evolved into a planetwide superorganism that, for the mystically inclined cleric, would coincide with the incarnation of the Christian God in the physical universe.[27]

Teilhard's overtly religious conception of the Omega Point is distinct from the secular version articulated by Barrow and Tipler in their book The Anthropic Cosmological Principle.[28] *For them, the Omega Point is the hypothesized final point in the evolution of a linked set of closed universes (a so-called multiverse) that proceed to contract toward a Big Crunch billions of years hence. Assuming that life persists and continues to evolve until the Omega Point, Barrow and Tipler*

- Why are we here?
- Why is there something rather than nothing?
- Are we alone?
- Was the cosmos designed?
- Who or what created the universe and endowed it with anthropic qualities?
- Is the universe evolving toward a discernible Point Omega—what Polanyi called "an unthinkable consummation"—in the distant future? (See sidebar, page 10.)
- What fate awaits our descendants at what may possibly be the end of time?
- What cosmic role might life itself play at Point Omega?
- Will Point Omega represent the final terminus of the cosmos—the concluding chord in a magnificent symphony—or might it conceivably constitute a threshold to another iteration of an endless process of cosmological evolution?
- Might the cosmos itself constitute neither a "one-shot affair" inaugurated by the Big Bang nor an infinitely branching chaotic fractal but rather a closed, timelike curve whose definitive characteristic is eternal, recursive self-organization?

Many courageous men and women from a great variety of scientific disciplines and a multitude of cultures have begun to focus serious attention on these daunting questions. Like those intrepid navigators Thor Heyerdahl and his companions who fearlessly launched the frail

project that life's counterentropic organizational power will then come to dominate inanimate matter and energy completely:

At the instant the Omega Point is reached, life will have gained control of all matter and forces not only in a single universe, but in all universes whose existence is logically possible; life will have spread into all spatial regions in all universes which could logically exist, and will have stored an infinite amount of information, including all bits of knowledge which it is logically possible to know. And this is the end.[29]

My usage of the term Point Omega *corresponds to the Barrow and Tipler definition. My key difference with these two authors is that I disagree with the final sentence quoted above.*

Kontiki raft into the dark waters of the Pacific Ocean with no definite plan for their voyage and an insouciant disregard for the possibility of failure, these brave pioneers of science have commenced an astounding voyage of discovery that many of their peers will dismiss as quixotic. Their daring mission is the stuff of myth: to tease out the final secrets lurking in the cosmic code and, in the process, to uncover the great artist's hidden signature on the vast unfolding canvas we call the universe.

Chapter Two

A Life-Friendly Cosmos

The Idea of Nothing

John D. Barrow is a brilliant polymath who as of this writing serves as a research professor of mathematical sciences at Cambridge University. A colleague of Stephen Hawking, Barrow has written a number of bestselling popular books about cosmology and the future direction of science.

In The Book of Nothing, *published in 2001, Barrow recounts the history of an idea that is at the heart of both mathematics and modern conceptions of the nature of the universe: nothingness. Barrow begins by tracing the origin of the concept of zero to ancient India and then recounts its rocky reception and eventual acceptance in Europe.[2] As he puts it,*

(continued on next page)

In the beginning was the void—a nothingness so perfect and sublime that neither time nor space nor energy nor any material thing blemished its immaculate nonexistence. Then, by means of a mysterious force whose precise nature may forever elude what Charles Darwin called humankind's "godlike intellect,"[1] the universe was born. In a fleeting spasm we have come to know as the Big Bang, the cosmos leapt into violent infancy, emerging miraculously from the loins of nothing at all. (See sidebar page 13.)

From that abrupt and singular beginning flowed forth all the radiation and all the constituents of atoms that now reside (albeit in altered form) in galaxies, stars, planets, and other configurations (including human beings) in the great celestial bestiary. But something else—something genuinely extraordinary—emerged from the initial cataclysm as well: all the laws and constants of physics that govern the organization of all matter and energy throughout the universe. And therein lies the greatest of all wonders—a mystery captured memorably by Einstein's famous statement that what really interested him was whether God had any choice in designing the laws of nature. The impression of design in the laws of nature is unmistakable and utterly compelling. Einstein himself once confessed to a "rapturous amazement at the harmony of natural law, which reveals an intelligence of such superiority that, compared with it, all the systematic thinking and acting of human beings is an utterly insignificant reflection."[9]

If the full suite of natural laws and physical constants that emerged from the Big Bang had been received in the form of an intercepted radio signal from Alpha Centauri (one of the stars nearest Earth), the champagne corks would surely be popping in the hallways of the SETI Institute, the informal world headquarters for the search for extraterrestrial intelligence. But because these laws and constants are embedded in the very fabric of our reality—stitched like the finest embroidery into the essence of our existence—most of us tend to ignore the apparent signature of transcendent intelligence. However, there are exceptional individuals, like Albert Einstein and Heinz Pagels, who have striven to grasp the deeper implications of the perplexing mystery of a meticulously structured physical universe, adhering with unreasonable precision to beautifully ordered and intricately interrelated mathematical principles.

For these iconoclasts, the very order of the cosmos is itself a vital clue about the existence of some fundamental physical process that we do not yet comprehend. For them, the impression of design plays precisely the same role that it did for Charles Darwin 150 years ago: it serves not as a call to surrender to religious faith but rather as a subtle hint of deep natural principles waiting to be teased out by the enlightened minds of scientists as well as an irresistible goad to further inquiry.

The analogy to Darwin and the theory of evolution is actually even more precise. The truly amazing feature about the intricate order of the cosmos is that it appears to be exquisitely fine-tuned to permit the existence of life-forms like ourselves. As the renowned Princeton physicist John Wheeler wrote in a foreword to a seminal book by astrophysicists John Barrow and Frank Tipler entitled *The Anthropic Cosmological Principle*,

> It is not only that man is adapted to the universe. The universe is adapted to man. Imagine a universe in which one or another of the fundamental dimensionless constants of physics is altered by a few percent one way or the other? Man could never come into being in such a universe. That is the central point of the anthropic principle. According to the principle, a life-giving factor lies at the center of the whole machinery and design of the world.[10]

"Medieval science and theology grappled constantly with the idea of the vacuum, trying to decide questions about its physical reality, its logical possibility and its theological desirability."[3]

The biggest challenge in grasping the notion of nothingness is that it is the precise opposite of everyday experience, which is a constant stream of something—something perceived, something felt, something accomplished. In Barrow's words, the idea of zero or nonexistence seems to "invite paradox and confusing self-reference."[4]

Yet no idea is more central to modern theories of cosmology and physics. The Big Bang hypothesis is a theory of the origin of the universe ex nihilo. Einstein's theory of gravity as the curvature of the space-time continuum allows scientists to "describe a space that is empty of mass and energy with complete mathematical precision. Empty universes [can] exist."[5] And quantum mechanics provides a new conception of a vacuum state that is not truly empty but that represents the "lowest energy state available" and that can be mysteriously provoked into a higher energy state by any slight disturbance.[6]

Finally, as Barrow notes, the idea of a ghostly form of vacuum energy called "lambda" is the centerpiece of astrophysicists' attempt to explain a puzzling acceleration in the expansion rate of the universe. The strange lambda force—literally antigravity—seems to operate in odd and counterintuitive ways. For instance, it does not seem to diminish with astronomical distance, in contrast to gravity.

As Barrows reveals, the delicate interplay between the attractive force of gravity and the repulsive force of lambda holds the key to the fate of the universe: Will it terminate in eventual recollapse (a kind of Big Bang in reverse) or eternal expansion, resulting in a state of the cosmos in the distant future that Barrow characterizes as a "sea of non-interacting, fairly structureless collections

(continued on next page)

Indeed, so fine-tuned are the laws of physics to favor life that it almost seems as if the universe and the life-forms that inhabit it have coevolved, like earthly creatures and the ecosystems they populate. (Cosmologist Lee Smolin recently estimated that the statistical likelihood of a universe that exhibits the requisite combination of fine-tuned fundamental constants necessary for our kind of life to exist is on the order of one part in 10^{220}—one improbable roll of the dice in a random sequence so monstrously vast that it far exceeds the number of atoms in the universe.[11])

One key example of the astonishing array of "just so" coincidences inherent in the physical characteristics of our universe—characteristics that render the cosmos peculiarly friendly to carbon-based life—is described by physicist Stephen Hawking in an essay published in 1996 entitled "Quantum Cosmology." (See sidebar, page 17.)

> The trouble with the hot big bang model is the trouble with all cosmology that has no theory of initial conditions: it has no predictive power. Because general relativity would break down at a singularity, anything could come out of the big bang. So why is the universe so homogeneous and isotropic on a large scale, yet has local irregularities such as galaxies and stars? Any why is the universe so close to the dividing line between collapsing again and expanding indefinitely? In order to be as close as we are now, the rate of expansion early on had to be chosen fantastically accurately. If the rate of expansion one second after the big bang had been less by one part in 10^{10}, the universe would have collapsed after a few million years. If it had been greater by one part in 10^{10}, the universe would have been essentially empty after a few million years. In neither case would it have lasted long enough for life to develop. Thus one either has to appeal to the anthropic principle or find some physical explanation of why the universe is the way it is.[13]

The Anthropic Cosmological Principle: Historical Antecedents

Scientists who ponder mysteries concerning the fitness of the cosmic environment for life are exploring the ramifications of what is called the "anthropic cosmological principle," an idea that seeks to explain why the universe we inhabit is so oddly life-friendly. The idea that the universe is fine-tuned to support life has an ancient lineage. Anaxagoras, a pre-Socratic Greek philosopher, was among the first to conclude that the order and harmony of the universe betrayed evidence of some sort of design—the influence of a cosmic mind that induced the appearance of order from a primeval chaos. What is strikingly modern about the philosophy of Anaxagoras, particularly in contrast to that of his successor Aristotle, is the deistic notion that the cosmic mind that initially arranges the order of the cosmos thereafter departs the scene, never to be heard from again, and exerts no ongoing influence on cosmic evolution.

For the later philosophers Socrates and Plato, by contrast, the material universe was continually supported and ordered by supermaterial laws and forms that were accessible to enlightened minds. (See sidebar, page 19.)

Aristotle took this concept one step further and theorized that reality can only be adequately understood as matter acting in obedience to a series of nested causal factors he termed "efficient" causes (which correspond to the contemporary notion of causation, wherein the cause always precedes the effect in time) and "final" causes (the teleological notion that every entity or process is drawn irresistibly toward the purpose of that object or process). This idea, strongly reminiscent in some ways of the contemporary concept of emergence, posits that final causes are vastly more important than efficient causes in structuring a meticulously ordered cosmos. As Barrow and Tipler observe, "Aristotle's guiding principle was that the ultimate meaning of things was to be divined from their 'end' (*telos*) rather than their present configurations—that is, by learning of their final rather than their material causes."[16] Under Aristotle's vision, the evolution of the cosmos was purposefully drawn,

of stable elementary particles and radiation."[7] In other words, will the universe end in fire or ice? It is the precise value of the mysterious lambda force—the negative energy of the vacuum itself—that holds the answer to this profound question.

What does all of this have to do with life on planet Earth? Barrow points out that the value of lambda and its relationship to gravity not only foretells the ultimate fate of the universe but is also intimately related to the possibility of life's appearance:

Although huge values of lambda are the most probable and persistent, they give rise to a universe that expands too fast too early for stars and galaxies and astronomers ever to appear. If we were casting our eye across all possible universes displaying all possible values of the lambda stress, it could be that those, like our own, with outlandishly small values are self-selected from all the possibilities by the fact that they are the only ones that permit observers to evolve.[8]

Large Number Coincidences and the Nature of the Universe

The existence of very large numbers in quantitative descriptions of the cosmos should not be surprising. The cosmos is, after all, very large indeed in comparison to the realm of existence with which we are familiar. The vastness of cosmic numbers—the size of the galaxy measured in miles, for instance, or the distance to a giant quasar in meters—merely reflects this dizzying disparity of scale.

What is more interesting is a series of puzzling coincidences in the numbers that define the basic parameters of the universe. Reluctant to dismiss these coincidences as nothing more than numerological serendipity, imaginative theorists have hypothesized that there is a deep causal link between such seemingly unrelated large numbers as the ratio of the age of the universe to the time it takes a photon to traverse an atom

(continued on next page)

as if by a kind of sublime magnetism (a perfecting principle called "entelechy"), to a final state of perfection. The degree of harmony evident in nature at any particular historical stage corresponds to the degree of progress toward that final perfect state.

With the advent of modern science, the concept of teleological causation was largely supplanted by a new focus on mechanical (or efficient) causation. Indeed, the seventeenth-century philosopher Baruch Spinoza, whose thinking greatly inspired Albert Einstein, viewed the concept as a simple logical error—a regrettable conflation of the ideas of cause and effect:

> It remains to be shown that nature does not propose to itself any end in its operations, and that all final causes are nothing but pure fictions of human imagination. I shall have little trouble to demonstrate this; for it has already been firmly established. . . . I will, however, add a few words in order to accomplish the total ruin of final causes. The first fallacy is that instead of regarding as a cause that which is by nature anterior, it makes the cause posterior.[17]

Yet there remained stubbornly defended redoubts of belief in the ancient concept of teleological causation. Perhaps the most interesting example of adherence to this seemingly outdated notion was that of Isaac Newton, the father of mathematical physics and arguably the greatest scientist who ever lived. For Newton and at least some of his intellectual progeny, the magnificent order of the cosmos captured in his monumental *Principia* was irrefutable evidence of both teleological causation and supernatural design.

This aspect of Newton's thought, while largely ignored by contemporary scientists, became the inspiration for intelligent design theories of various sorts. Countering these theories was a set of ideas put forward by the Scottish philosopher David Hume that foreshadowed the concepts later to be articulated by Charles Darwin. Hume's attacks on the possibility of an inference of design centered on the impossibility of genuine proof, the lack of dispassionate consideration of contrary evidence of chaos as a pervasive force in nature, and the inherent limitations imposed by nature on the human mind that prevent it from reaching such grand conclusions with any reasonable degree of certainty.

Hume did not merely oppose the idea of intelligent design but also counterposed a concept that at the time was viewed as eccentric and counterintuitive: that the universe is fundamentally organic rather than mechanical in nature and, as an organic being, is possessed of a mysterious self-organizational propensity. Dismissed as bizarre and out of the mainstream, Hume's vision of nature as an organic, self-ordering entity ultimately had enormous influence on Erasmus Darwin, Charles Darwin's grandfather, who took the first steps toward the scientific revolution for which his grandson was responsible. For Erasmus Darwin, the key insight supplied by Hume was that a self-organizing cosmos and biosphere would exhibit the phenomenon of "meliorism"—evolution that exhibited gradual improvement while progressing toward a final state of perfection intended by an intelligent designer:

> The late Mr. Hume . . . concludes that the world itself might have been generated, rather than created; that is, it might have been gradually produced from very small beginnings increasing by the activity of its inherent principles, rather than by a sudden evolution of the whole by the Almighty fiat. What a magnificent idea of the infinite power to cause the causes of effects, rather than to cause the effects themselves.[18]

Differing from this hybrid concept of design and evolution were the views of Charles Darwin's famous predecessor, William Paley. A distinguished Cambridge professor who will forever live in infamy in the annals of science as Darwin's intellectual foil, Paley offered up in his 1802 book *Natural Theology* the classic argument for intelligent design: the analogy of elaborately constructed living creatures to a finely crafted watch. As Paley put it, "In crossing a heath . . . suppose I had found a *watch* upon the ground, and it should be inquired how the watch happened to be in that place. . . . [I would be forced to conclude] that the

(approximately 10^{39}) and the ratio of the electric force between electrons and protons and the gravitational force between electrons and protons (which also—and rather mysteriously—just happens to be approximately 10^{39}).

The physicist P. A. M. Dirac offered a radical explanation for this startling coincidence in the form of his Large Numbers Hypothesis, which was "Any two of the very large dimensionless numbers occurring in Nature are connected by a simple mathematical relation, in which the coefficients are of the order of magnitude unity."[12]

Because Dirac's hypothesis incorporated numerical ratios that depend upon the age of the universe, a startling and eminently testable implication of his proposal is that the fundamental constants of nature (such as the gravitational constant) will vary with time. This hypothesis, which remains intensely controversial, inaugurated what is now a vast literature in astrophysics examining the possibility and the testability of variation over time in many of the so-called constants of nature, such as the fine-structure constant and even the speed of light.

watch must have had a maker: that there must have existed, at some time, and at some place or other, an artificer or artificers, who formed it for the purpose which we find it actually to answer; who comprehended its construction, and designed its use."[19] The existence of a fine watch displaying indisputable evidence of design implied, Paley concluded, the existence of a watchmaker. Therefore, the existence of a finely crafted order in living creatures implied the existence of a divine designer.

While Darwin's subsequently promulgated theory of evolution through natural selection dealt a death blow to Paley's naive argument in favor of the existence of a divine designer on the basis of biological design, another aspect of Paley's thought remains at least tangentially relevant to modern expressions of the cosmological anthropic principle. The second part of Paley's *Natural Theology*, which is generally overlooked because it has not been made the object of scorn by ultra-Darwinists like Oxford biologist Richard Dawkins, concerns evidence of design emanating from inorganic nature. While most of these arguments appear naive to modern scientists, one in particular foreshadows modern anthropic reasoning: the anthropic implications of the inverse square law of gravitation. This basic law, first fully articulated by Isaac Newton, provides that the force of gravity diminishes in accordance with the square of the distance separating two gravitating bodies (like Earth and the Moon). As Barrow and Tipler note,

> The next observations [Paley] makes are the most intriguing from a modern perspective: he points out the unique features that are intrinsic to Newton's inverse square law of gravitation. The basis for his comparative study is an imaginary ensemble containing all possible power laws of variation for the gravitational force. The size of the subset of this collection which are consistent with our existence can then be examined in Anthropic fashion.[20]

Paley reaches a conclusion in striking conformity with modern analyses of the anthropic implications of the fact that there are only

The Intellectual Patrimony of Plato

The concept articulated by Plato that the material world we experience is merely a pale reflection of an underlying reality of immaterial forms has enjoyed a remarkable longevity over the centuries. The notion recently reappeared as the foundation for a remarkable work of speculative physics by Julian Barbour entitled The End of Time: The Next Revolution in Physics.[14]

Barbour asks us to entertain the possibility that what we perceive as the flow of time is an illusion and that time does not truly exist as a fundamental feature of nature. Instead, ultimate reality consists of a "block universe" in which every

(continued on next page)

three extended spatial dimensions in our cosmos (out of a theoretically possible set of ten spatial dimensions predicted by contemporary M-theory, the notion that ultimate reality consists of tiny strings and membranes), yielding an inverse square law for the diminution of the force of gravity in three-dimensional space:

> Whilst the possible laws of variation were infinite, the admissible laws, or the laws compatible with the preservation of the system, lie within narrow limits. If the attracting force [of gravity] had varied according to any direct law of the distance, let it have been what it would, great destruction and confusion would have taken place. The direct simple proportion of the distance would, it is true, have produced an ellipse [in the form of a planetary orbit]; but the perturbing forces would have acted with so much advantage, as to be continually changing the dimensions of the ellipse, in a manner inconsistent with our terrestrial creation.[21]

Likewise, if gravity had diminished in our cosmos in accordance with an inverse *cube* law (or indeed any inverse power law by which the force of gravity diminishes more rapidly than under the dictates of the inverse square), life as we know it could not exist on planetary surfaces. Says Paley,

> Of the inverse laws, if the centripetal force had changed as the cube of the distance, or in any higher proportions . . . the consequence would have been, that the planets, if they once began to approach the sun, would have fallen into its body; if they once, though by ever so little, increased their distance from the centre, would forever have receded from it. . . . All direct ratios of the distance are excluded, on account of the danger from perturbing forces; all reciprocal ratios, except what lies beneath the cube of the distance . . . would have been fatal to the repose and order of the system. . . . The permanency of our [planetary] ellipse is a question of life and death to our whole sensitive world.[22]

instant—past, present, and future—coexists simultaneously.

In deference to the abiding power of Plato's original vision, Barbour calls this mathematically perfect and eternal state Platonia—a "timeless landscape" where nothing changes and where all the illusory instants of time, "all the Nows," are "simply there, given once and for all."[15]

Karl Popper and Falsifiability

Karl Popper was a philosopher of natural and social science who formulated what has come to be known as the falsifiability criterion for distinguishing genuine science from such nonscientific disciplines as metaphysics and religion. Popper's concept of falsifiability as the hallmark of scientific thinking is summarized as follows on the official Karl Popper website:

Falsificationism is the idea that science advances by unjustified, exaggerated guesses followed by unstinting criticism. Only hypotheses capable of clashing with observation reports are allowed to count as scientific. "Gold is soluble in hydrochloric acid" is scientific (though false); "Some homeo-pathic medicine does work" is, taken on its own, unscientific (though possibly true). The first is scientific because we can eliminate it if it is false; the second is unscientific because even if it were false we could not get rid of it by confronting it with an observation report that contradicted it. Unfalsifiable theories are like the computer programs with no uninstall option that just clog up the computer's precious storage space. Falsifiable theories, on the other hand, enhance our control over error while expanding the richness of what we can say about the world.[50]

Some purists question whether such esoteric

(continued on next page)

This striking anthropic coincidence—that gravity diminishes in accordance with the only physical law that is consistent with the existence of life on planetary surfaces like Earth—was, for Paley, "perhaps the strongest evidence of care and foresight that can be given."[23]

While the publication of Darwin's theory sounded the death knell for Paley's argument from design on the basis of biological complexity, the theory of evolution through natural selection did not purport to address or explain the curious astrophysical coincidence noted by Paley: that the inverse square law of gravity is uniquely life-friendly. It is a measure of the level of disrepute to which Paley's thinking fell after publication of *The Origin of Species* that this observation has simply been overlooked.

Modern Statements of the Cosmological Anthropic Principle

Modern statements of the cosmological anthropic principle date from the publication of a landmark book by Harvard professor Lawrence J. Henderson in 1913 entitled *The Fitness of the Environment.*[24] Henderson's book was an extended reflection on the curious fact that there were particular substances present in the environment—preeminently water—whose peculiar qualities rendered the environment almost preternaturally suitable for the origin, maintenance, and evolution of organic life.[25] Indeed, the strangely life-friendly qualities of these materials led Henderson to the view that "we were obliged to regard this collocation of properties in some intelligible sense a preparation for the process of planetary evolution. . . . Therefore the properties of the elements must for the present be regarded as possessing a teleological character."[26]

Thoroughly modern in outlook, Henderson dismissed this apparent evidence that inanimate nature exhibited a teleological character as indicative of divine design or purpose. Indeed, he rejected the notion that nature's seemingly teleological quality was in any way inconsistent with Darwin's theory of evolution. On the contrary, he viewed the biofriendly character of the inanimate natural environment as absolutely essential to the optimal operation of the evolutionary forces in the biosphere. Absent the substrate of a superbly "fit" inanimate environment, Darwinian evolution could never have achieved what it has in terms of species multiplication and diversification.

The great mystery of *why* the physical qualities of the inanimate universe just happened to be so oddly conducive to life and biological evolution remained just that for Henderson—an impenetrable mystery. The best he could do to solve the puzzle was to speculate that the laws of chemistry were somehow fine-tuned in advance by some inconceivable cosmic evolutionary mechanism to meet the future needs of a living biosphere:

> The properties of matter and the course of cosmic evolution are now seen to be intimately related to the structure of the living being and to its activities; they become, therefore, far more important in biology than has previously been suspected. For the whole evolutionary process, both cosmic and organic, is one, and the biologist may now rightly regard the Universe in its very essence as biocentric.[27]

Henderson's brave and iconoclastic vision was, alas, far ahead of its time. His potentially revolutionary book was largely ignored by his contemporaries or dismissed as a mere tautology. *Of course* there should be a close match-up between the physical requirements of life and the physical world that life inhabits, contemporary skeptics pointed out, since life evolved to survive the very challenges of that pre-organic world and to take advantage of the biochemical opportunities it offered.

While lacking broad influence at the time, Henderson's pioneering vision proved to be the precursor to modern formulations of the cosmological anthropic principle. These formulations, which vary greatly in sophistication and persuasiveness, draw upon contemporary interpretations of puzzling physical phenomena. A few examples among many others are discussed below.

mathematical disciplines as M-theory qualify as genuine science. Why? Because the minuscule microworld that M-theory seeks to explain (an eleven-dimensional realm of vibrating strings and membranes far tinier than the smallest subatomic particle) is probably impossible to probe with earthly instrumentation. M-theorists are now seeking to remedy this problem by searching for the signature of hidden dimensions in the cosmic microwave background radiation. That background is the faint echo of the greatest and most powerful experiment in nuclear physics that likely ever took place: the Big Bang.

The Anthropic Miracle of Carbon Nucleosynthesis: Fred Hoyle's Lasting Legacy

British astronomer Fred Hoyle was responsible for one of the great astrophysical blunders of the twentieth century: the notion that the cosmos was in a "steady state" of constant expansion, accompanied by the unvarying creation of new matter in empty space. This colossal error was premised on the seeming absurdity of the notion that our vast universe could have had a definite beginning in the far distant past.

Hoyle was not alone in this view. Einstein was repelled by the notion that his own theories seemed to suggest that the cosmos originated in a singularity long ago. To compensate for this unfortunate implication of the theory of relativity, Einstein concocted the notion of a cosmological constant that acted like a kind of antigravity and ensured that the universe need not have been born in such a manner. Einstein later conceded that this idea was the worst mistake he had ever made (although it now appears that Einstein was indeed correct about the possible existence of such an antigravitational force).[28]

When the evidence began to accumulate that the cosmos had, in fact, come into existence in a fantastic cataclysm long ago—an event that Hoyle derisively termed a "Big Bang" (a term that has become, somewhat ironically, the standard shorthand phrase for referring to the event that theoretically commenced the beginning of time and the evolution of the physical universe)—Hoyle's reputation as a visionary cosmologist suffered immeasurably. What should not be forgotten, however, is that Hoyle had one other very substantial claim to visionary status that survives to this day: the first successful scientific prediction ever to result from anthropic reasoning. Hoyle was one of the first scientists to conclude that all the heavy elements—which is to say all the elements other than the handful of extremely light substances—are the product of monstrously powerful thermonuclear reactions that take place in the interiors of giant stars. Only the lightest elements, principally hydrogen and helium, emerged directly from the residue of the Big Bang.

Among the heavy elements to emerge from the process of stellar nucleosynthesis is carbon, which is essential for life. The carbon in our bodies is the product of this nuclear alchemy—the residue of titanic supernovae that exploded long ago, spewing their life-giving residue throughout our galaxy. As Carl Sagan once observed with complete accuracy, we are star-stuff, the ashes of ancient stellar holocausts.

What Hoyle realized was that in order for carbon to exist in the abundant quantities that we observe throughout the cosmos, the mechanism of stellar nucleosynthesis must be exquisitely fine-tuned in a very special way. In essence, Hoyle hypothesized, the process of thermonuclear fusion inside of stars that yields carbon atoms must benefit from a kind of built-in booster mechanism that catalyzes a two-step process by which two helium atoms first fuse together to form an unstable intermediate element (beryllium) and then another helium nucleus is captured by the beryllium atom to form carbon. Because this two-step process is inherently precarious and improbable, it will take place on a routine basis only if the nuclear energies of the various colliding atoms are tuned with extraordinary precision so that the beryllium atom is "sticky" to the third fusing helium atom. Hoyle predicted that the "stickiness" would inhere in hypothesized matching nuclear resonance levels in the carbon atom and that of prospective merger candidates (beryllium and helium). Such a match seemed unlikely in the extreme on the basis of theoretical nuclear physics, but, in fact, subsequent experiments triumphantly confirmed Hoyle's predictions.

But the miraculous story of carbon nucleosynthesis does not end there. In order for carbon atoms to persist in sufficient quantity throughout the extraordinarily violent process of stellar nucleosynthesis, it must not be too easy for carbon to be transmuted into the next element in the periodic table (oxygen). Indeed, there must be a precisely engineered "speed bump" in the transmutation process that impedes the synthesis of oxygen from carbon. As it turns out, this barrier also exists, built into the very fabric of subatomic reality. Without plentiful carbon, there would be no carbon-based life. Without the exquisite fine-tuning at the level of elementary physics, the universe would lack plentiful supplies of life-giving carbon. Hoyle's uncanny predictions about carbon nucleosynthesis have been widely and properly hailed as an important precedent demonstrating that anthropic reasoning can be truly scientific in the sense of generating falsifiable implications.

Martin Rees: The Magic Number of the Anthropic Cosmos Is Six

Martin Rees is the Astronomer Royal of England, a medieval-sounding position that provides him with a unique "bully pulpit" to help set the agenda for space science and cosmological research throughout the world. Because of Rees's unique eminence, his remarkable book published in 2000—*Just Six Numbers: The Deep Forces That Shape the Universe*—is helping to transform investigations into the nature and possible origin of anthropic coincidences built into the laws of physics into mainstream science.[29]

Indeed, *Just Six Numbers* is a kind of intellectual agenda for cosmologists interested in investigating the nuances of the anthropic cosmological principle. His book is an extended reflection on the astonishing fact that every aspect of the evolution of the universe—from the birth of galaxies to the origin of life on Earth—is sensitively dependent on the precise values of seemingly arbitrary constants of nature like the strength of gravity, the number of extended spatial dimensions in our universe, and the initial expansion speed of the cosmos following the Big Bang. If any of these physical constants had been even slightly different, life as we know it would have been impossible. As Rees puts it,

> The [cosmological] picture that emerges—a map in time as well as in space—is not what most of us expected. It offers a new perspective on a how a single "genesis event" created billions of galaxies, black holes, stars and planets, and how atoms have been assembled—here on Earth, and perhaps on other worlds—into living beings intricate enough to ponder their origins. There are deep connections between stars and atoms, between the cosmos and the microworld. . . . Our emergence

> and survival depend on very special "tuning" of the cosmos—a cosmos that may be even vaster than the universe that we can actually see.[30]

Let us consider each of the six key numbers briefly in order to appreciate their anthropic significance.

N: The Ratio of the Electrical Force to the Gravitational Force

Both gravity and the force of electrical attraction are governed by the inverse square law (that is, both forces diminish with the square of the distance separating two attractive bodies) but there the similarity, for all practical purposes, ends. Electrical attraction is enormously strong while gravity is relatively weak. The number *N*—which defines the ratio between electrical attraction (the he-man force) and gravity (the ninety-pound weakling)—is 10^{36}, which is the number one followed by thirty-six zeros. This is an extraordinarily large number for which we have no common name (like trillion or quadrillion) and yet that is utterly essential to the existence in the cosmos of our kind of life.

Why? Because if gravity were only a tad bit stronger than it actually is, all sorts of events injurious to the evolution of life would occur. Here is how Rees describes the adverse consequences for living creatures of a cosmos in which the relative strength of gravity was increased just slightly:

> Galaxies would form much more quickly in such a universe, and would be miniaturized. Instead of the stars being widely dispersed, they would be so densely packed that close encounters would be frequent. This would in itself preclude stable planetary systems, because the orbits would be disturbed by passing stars—something that (fortunately for our Earth) is unlikely to happen in our own Solar System. But what would preclude a complex ecosystem even more would be the limited time available for development. Heat would leak more quickly from these "mini-stars": in this hypothetical strong-gravity world, stellar lifetimes would be a million times shorter. Instead of living for ten billion years, a typical star would live for about 10,000 years. A mini-

> Sun would burn faster, and would have exhausted its energy before even the first steps in organic evolution had got under way. Conditions for complex evolution would undoubtedly be less favourable if (leaving everything else unchanged) gravity were stronger.[31]

ε: The Strength of the Force Binding Protons and Neutrons in an Atomic Nucleus

Galaxies are vast ecosystems, recycling the residue of exploded stars into giant clouds of cosmic debris available for recycling in successive stellar generations and new planets like Earth. This massive recycling process is essential for the existence of life as we know it because virtually none of the myriad elements upon which life depends (like carbon and oxygen) would have been produced in the original cataclysm known as the Big Bang. These elements are instead the fruits of a very special kind of alchemy that occurs only during the violent death throes of giant stars far larger than our sun.

In medieval times, the magical source of the alchemist's power—the secret tool that empowered him to transmute lead into gold—was known as the philosopher's stone. The question now is, What is the modern counterpart to the philosopher's stone—the secret ingredient that allows the mysterious process of stellar alchemy to proceed in a life-friendly manner, yielding plentiful quantities of precisely those elements needed for a carbon-based metabolism and only trace amounts of the heaviest radioactive elements like uranium?

It is no sorcerer's wand but something far more wondrous that accounts for this apparent miracle: a precise adjustment on one of nature's tiniest dials. The meticulous tuning of ε—the number that defines the exact strength of the so-called strong nuclear force that binds protons and neutrons together in atomic nuclei—is nature's equivalent of the mythical philosopher's stone. It is this number, equal approximately to .007, that accounts for the ability of nature to fuse together the primitive elements spewed forth in copious quantities in the Big Bang and create the entire periodic table of elements well known to every high school student of chemistry.

The strong nuclear force seems incredibly remote from our daily lives. It operates, after all, only at the most minute scale of nature (inside atomic nuclei) and diminishes so rapidly that even subatomic particles within large nuclei (like those of the uranium atom) are able to marginally evade its grasp. But so stupendously powerful is this small-scale force—so massively greater is it than the power of electromagnetism or gravity—that it alone, of all nature's basic forces, is sufficient to fuel the burning of our sun for billions of years.

The power of this force is overwhelming. It is not only the force that gives us the gift of sunshine but also the source of the horrendous power that leveled Hiroshima and Nagasaki. It is the force that drives temperatures in the hearts of supernovae up to a billion degrees, transforming them into crucibles in which all the higher elements are forged. And it is a force that is, in the precise magnitude at which it exists in our universe, absolutely essential to the origin and maintenance of carbon-based life. If the strong force had been marginally weaker, even by a tiny amount, the results would have been disastrous for the future of life. As Martin Rees puts it,

> The crucial first length in the chain of [nucleosynthesis]—the build-up of helium from hydrogen—depends rather sensitively on the strength of the nuclear "strong interaction" force. A helium nucleus contains two protons, but it also contains two neutrons. Rather than the four particles being assembled in one go, a helium nucleus is built up in stages, via deuterium (heavy hydrogen), which comprises a proton plus a neutron. If the nuclear "glue" were weaker, so that ε were 0.006 rather than 0.007, a proton could not be bonded to a neutron and deuterium would not be stable. Then the path to helium formation would be closed off. We would have a simple universe composed of hydrogen, whose atom consists of one proton orbited by a single electron, and no chemistry. Stars could still form in such a universe (if everything else were kept unchanged) but they would have no nuclear fuel. They would deflate and cool, ending up as dead remnants. There would be no explosions to spray the debris back into space so that new

> stars could form from it, and no elements would exist that could ever form rocky planets.[32]

But what if the strong nuclear force were slightly more powerful? Would this not render the universe even more life-friendly than it actually is? It would not, as Rees reveals:

> At first sight, one might have guessed from this reasoning that an even stronger nuclear force would have been advantageous for life, by making nuclear fusion more efficient. But we wouldn't have existed if ε had been more than 0.008, because no hydrogen would have survived from the Big Bang. In our actual universe, two protons repel each other so strongly that the nuclear "strong interaction" force can't bind them together without the aid of one or more neutrons (which add to the nuclear "glue," but, being uncharged, exert no extra electrical repulsion). If ε were to have been [more than] 0.008, then two protons would have been able to bind directly together. This would have happened readily in the early universe [immediately after the Big Bang], so that no hydrogen would remain to provide the fuel in ordinary stars, and water would never have existed.[33]

Nor, for that matter, would carbon or oxygen or any of the higher elements of nature that are so necessary for life have ever had a chance to come into existence in the furnaces of cosmic transmutation at the heart of giant supernovae.

Ω: The Fine-Tuning of the Rate of Cosmic Expansion

Our cosmos is vast beyond the human power of imagination. It encompasses billions upon billions of giant galaxies, each composed of millions to billions of stars and each separated by thousands to billions of light-years (the distance that light travels in empty space in one year). Yet the immense size of the universe, so far beyond the scale of human life, is absolutely essential to the existence of carbon-based biology on Earth. The reason is that, just as is the case with a living creature growing to maturity, size equates roughly to age. What is critical for the existence of life is that the universe be sufficiently *old* that sufficient time will have elapsed since the Big Bang for the first generations of stars to be born and then die, spewing forth the life-giving elements like carbon into interstellar space. These elements are then incorporated into subsequent generations of stars and planets and eventually are available for use in constructing the complex pathways of prebiotic chemistry, which is the platform for life itself. After life begins, sufficient time must be available for the processes of organic evolution to work their magic before anything like human-level or higher intelligence can emerge on any planetary surface.

Simply put, if the universe were not so large it could not possibly be so old, and if it were not so old, there could be no life or intelligence on Earth or any other planet. What, then, is the mystery that explains the antiquity of the universe? Simply this: the exquisite balance that our universe has achieved between its rate of expansion and the force of gravity pulling all of its constituent elements—stars, planets, clouds of interstellar dust, and mysterious "dark matter"—back together.

The exact value of the number that defines a theoretical perfect balance between these two cosmic forces—the outward rate of expansion and the inward pull of gravity—is known to cosmologists as the "critical density." The ratio of the actual density that our universe exhibits and the theoretical critical density is known as omega (Ω). This is arguably the most important number in the universe. Why? Because if Ω is exactly equal to one, then the universe will continue expanding indefinitely but at a uniform rate. If Ω is slightly less than one, the universe will also expand indefinitely but at an ever increasing rate, yielding an essentially empty cosmos in the far distant future. (This is known as an "open universe" scenario.) On the other hand, if Ω turns out to be ever so slightly more than one, the expansion of the universe will eventually slow down, halt, and then reverse as the universe begins to contract toward a Big Crunch. (This is known as a "closed universe" scenario.)

What is more immediately important for the possibility of life than the ultimate fate of the universe is the intermediate effect of the value of

Ω on the evolution of the universe from the moment of the Big Bang up to and including the present cosmic era. Martin Rees explains why this effect is so critical:

> There are . . . good grounds for extrapolating back to when the universe was one second old and at a temperature of ten billion degrees. Suppose that you were "setting up" a universe then. The trajectory it would follow would depend on the impetus it was given. If it were started *too fast*, then the expansion energy would, early on, have become so dominant (in other words, Ω would have become so small) that galaxies and stars would never have been able to pull themselves together via gravity, and condense out; the universe would expand forever, but there would be no chance for life. On the other hand, the expansion must not have been too slow: otherwise the universe would have recollapsed too quickly to a Big Crunch.[34]

The astonishing degree to which Ω converged on the number one (a quantity that cosmologists refer to as "unity") in the very earliest cosmic era is perhaps the most convincing evidence of all that our universe is exquisitely fine-tuned to allow the possibility of life. As Rees puts it,

> Any emergent complexity [in the early universe] must feed on non-uniformities in density and in temperature (our own biosphere, for example, energizes itself by absorbing the Sun's "hot" radiation and re-emitting it into cold interstellar space). Without being to the slightest degree anthropocentric in our concept of life, we can therefore conclude that a universe has to expand out of its "fireball" state, and at least cool down below 3000 degrees, before any life can begin. If the initial expansion were too slow to permit this, there would be no chance for life. In this perspective, it looks surprising that our universe was initiated with a very finely-tuned impetus, almost exactly enough to balance the decelerating tendency of gravity. It's like sitting at the bottom of a well and throwing a stone up so that it just comes to a halt exactly at the top.[35]

This is no ordinary "well," of course, but one that is billions upon billions of light-years "deep" (the depth of this gravity well is equal to the distance that light has traveled from the moment of the Big Bang to the moment you are reading this sentence). Accordingly, the degree of precision required in "throwing" the stone of creation out of the gravity well of the initial singularity with exactly the right force to create a fine-tuned rate of cosmic expansion staggers the imagination. Here is Rees again on the truly miraculous character of this signature example of the anthropic qualities of our cosmos:

> The required precision is astonishing: at one second after the Big Bang, Ω cannot have differed from unity by more than one part in a million billion (one in 10^{15}) in order that the universe should now, after ten billion years, be still expanding and with a value of Ω that has certainly not departed wildly from unity. . . . Only a universe with a "finely tuned" expansion rate can provide the arena for [crucial life-giving nuclear and chemical] processes to unfold. So Ω must be added to our list of crucial numbers. It had to be tuned amazingly close to unity in the early universe. If expansion was too fast, gravity could never pull regions together to make stars or galaxies; if the initial impetus were insufficient, a premature Big Crunch would quench evolution when it had barely begun.[36]

λ: Antigravity: The Faintest and Most Mysterious Force in the Cosmos

The purported ability of advanced extraterrestrial civilizations to reverse the force of gravity and use antigravity as a means of achieving effortless spaceship propulsion is a staple of science fiction. But what if it were discovered that gravity could actually be nullified and reversed in the real world? Would this not be front-page news in every major publication in the world (and not just supermarket tabloids)?

As it turns out, precisely such a discovery was made a few years ago, not by a secret government agency pursuing an exotic new propulsion system for advanced spacecraft but in the context of the delicate task of measuring the distance to supernovae on the very edge of the visible universe. The amazing conclusion of the astronomy teams involved in this measurement effort was that the rate of expansion of space seemed to be *accelerating*, not slowing down, at the far frontiers of the universe. This meant, astrophysicists concluded, that a totally unknown force of cosmic repulsion—antigravity, if you will—begins to play a role at the most extreme distances accessible to our astronomical instruments.

This is a force about which we know precious little. It is far weaker than gravity, heretofore thought to be the weakest of the fundamental forces. It comes into its own only at vast distances defined by billions of light-years. And yet, over sufficient time, it may be able to overwhelm the attractive force of gravitation.

The one thing we know about λ (called lambda) is that it must be tiny in comparison with all the other forces of nature (including gravity) or it would long ago have emptied the cosmos of stars and planets and every material thing. Yet the mystery remains: How come λ is so incredibly tiny and yet not equal to zero? All that can be said for certain at this point is that the smallness of λ is unquestionably an anthropic constraint on our particular cosmos. As Martin Rees explains,

> A much higher value of λ would have had catastrophic consequences: instead of becoming competitive with gravity only after galaxies have formed, a higher-valued λ would have overwhelmed gravity earlier on, during the higher-density stages [of cosmic evolution]. If λ started to dominate before galaxies had condensed out from the expanding universe, or if it provided a repulsion strong enough to disrupt them, then there would be no galaxies. Our existence requires that λ should not have been too large.[37]

Q: Measuring the Wrinkles in Primordial Space

If the primordial fireball that emerged from the Big Bang had been completely uniform and featureless, then galaxies and stars could never have formed. Current cosmological theories surmise that these immense structures were seeded by tiny irregularities present at the initial stage of cosmic expansion. Cosmic inflation then expanded these wrinkles to immense size, allowing gravity to subsequently make use of them as traps to attract infalling matter, thus yielding all the subsequent structure of the universe.

But, as is the case with other seemingly magical numbers that define the anthropic qualities of our cosmos, the degree of primordial wrinkling had to be just right. If the original plasma were too smooth, no galactic structures could later have evolved; the universe would then have remained, for all of eternity, a boring soup of evenly dispersed particles and atoms. On the other hand, if the primordial plasma had been just a tiny bit too lumpy, the cosmos would not appear (as it does to us today) as nearly uniform on the largest scales. The precise degree of primordial lumpiness that obtains in our universe, quantified by the symbol *Q*, appears to be a prerequisite for a life-friendly cosmos. As Rees states, if the early universe had been just slightly less lumpy, life could never have subsequently evolved:

> If Q were *smaller* than 10^{-5} but the other cosmic numbers were unchanged, aggregations in the dark matter would take longer to develop and would be smaller and looser. The resultant galaxies would be anemic structures, in which star formation would be slow and inefficient, and "processed" material would be blown out of the galaxy rather than being recycled into new stars that could form planetary systems. If Q were smaller than 10^{-6}, gas would never condense into gravitationally bound structures at all, and such a universe would remain forever dark and featureless, even if its initial "mix" of atoms, dark matter and radiation were the same as in our own.[38]

On the other hand, if the primordial fireball that emerged from the Big Bang had been just slightly *more* lumpy than it actually was, the prospects for the emergence of life would have been equally bleak:

> A universe where Q were substantially *larger* than 10^{-5}—where the initial "ripples" were replaced by large-amplitude waves—would be a turbulent and violent place. Regions far bigger than galaxies would condense early in its history. They wouldn't fragment into stars but would instead collapse into vast black holes, each much heavier than an entire cluster of galaxies in our universe. Any surviving gas would get so hot that it would emit intense X-rays and gamma rays. Galaxies (even if they managed to form) would be much more tightly bound than the actual galaxies in our universe. Stars would be packed too close together and buffeted too frequently to retain stable planetary systems.[39]

Was the precise "lumpiness" of the primordial dough out of which nature kneaded galaxies and stars and ultimately living creatures somehow foreordained by the basic laws of physics? Or was it just a happy accident? As Rees points out, there is no reason to assume that any particular kind of spatial texture would be favored or disfavored in the early universe:

> The ripples would have been imprinted very early on, before the universe "knew" about galaxies and clusters; there would be nothing special about these sizes (or, indeed, about any dimensions that seemed significant in our present universe). The simplest guess would be that nothing in the early universe favors one scale rather than another, so that the ripples are the same on every scale. The degree of initial "roughness" was somehow established when our entire universe was of microscopic size.[40]

Long, long ago and far, far away, a microscope pot of primordial plasma, far tinier than the smallest atom, was stirred in precise accordance with the exacting recipe for a life-friendly cosmos—not too smooth, not too lumpy, but just right to eventually yield a magnificent diadem of stars and galaxies stretching across an immense void, with exactly the proper density and distribution to produce life-giving stars and terrestrial planets. How and why this miraculous fine-tuning of the very texture of space might have occurred is one of the abiding mysteries of modern cosmology.

D: Why Just Three Extended Spatial Dimensions?

We previously noted the anthropic consequence of the fact that the force of gravity diminishes in accordance with the inverse square law. This law is itself a secondary consequence of the fact that space in our universe possesses exactly three extended dimensions. If there were four extended spatial dimensions, gravity would diminish in accordance with an inverse cube law. If space had only two extended dimensions, gravity would diminish in direct proportion to the distance between two gravitating objects.

The serious study of the possibility that there might actually be more than three spatial dimensions is the province of "superstring" theory, or "M-theory" as it has recently come to be known. As Stephen Hawking recently noted, perhaps the biggest surprise for scientists probing the ultimate nature of reality is the increasing plausibility of the hypothesis that elementary particles like electrons and photons may actually be minuscule one-dimensional superstrings vibrating with delicate precision in the interstices of eleven-dimensional hyperspace. This bizarre notion seems to offer the best hope for unifying the seemingly incommensurable descriptions of nature provided by Einstein's theory of general relativity (which explains the behavior of space and time at the astrophysical scale) and quantum mechanics (which explains subatomic phenomena with exceptional accuracy).

The eleven-dimensional realm described by M-theory is inconceivably distant from everyday human experience. The tiny scale at which the theory describes nature will probably be forever inaccessible to even the most powerful atom smashers on Earth. Does this mean that M-theory is not science at all in the traditional sense? Hawking's answer is intriguing: "I must say that personally, I have been reluctant to believe in extra dimensions. But as I am a positivist, the question 'Do extra dimensions really exist?' has no meaning. All one can ask is whether mathematical models with extra dimensions provide a good description of the universe."[41]

And does M-theory offer a good description of the universe? Yes and no. While it is perhaps the best pathway currently available for approaching the Holy Grail of physics—a theory of quantum gravity—M-theory suffers from one perplexing weakness: it is capable of describing not only our universe but a whole menagerie of universes with varying constants and physical laws. How is it that our particular universe was favored to emerge in the Big Bang from this enormous multiplicity of possible universes?

According to Hawking, the answer may lie in the anthropic principle. This principle appears to Hawking (and many others) to be essential if we are to reconcile the nearly limitless range of alternative realities allowed by M-theory with the tightly constrained set of physical rules that govern the cosmos we actually inhabit. As Hawking puts it,

> The anthropic principle can be given a precise formulation, and it seems to be essential when dealing with the origin of the universe. M-theory . . . allows a very large number of possible histories for the universe. Most of these histories are not suitable for the development of intelligent life; either they are empty, last for too short a time, are too highly curved, or wrong in some other way. Yet according to Richard Feynman's idea of multiple histories, these uninhabited histories can have quite a high probability.[42]

This brings us back to the sixth anthropic question posed by Martin Rees: Why do we live in a universe with exactly three rather than two or even eight extended spatial dimensions? Is it because life as we know it could not otherwise exist? We have already noted the destabilizing effects on planetary orbits that would result from four or more extended spatial dimensions. What if there were fewer than three extended dimensions of space? Could life still exist in a two-dimensional "flatland" or even one-dimensional "lineland"? Here is Rees's tentative answer:

> Could we then live in a world where there were *less* than three [extended spatial dimensions]? The best argument here is a very simple one: there are inherent limitations on complex structures in "flatland" (or, indeed, on any two-dimensional surface). It is impossible to have a complicated network without the wires crossing; nor can an object have a channel through it (a digestive tract, for instance) without dividing into two. And the scope is still more constricted in a one-dimensional "lineland."[43]

Asking the question of whether the laws of nature dictate that precisely three spatial dimensions should have been extended following the Big Bang takes us to the very limits of contemporary scientific speculation. All we can say for certain at this stage is that the universe appears to be life-friendly (at least for life as we know it) if and only if there are exactly three extended dimensions of space and one dimension of time.

The Four Principal Versions of the Anthropic Cosmological Principle

The intriguing set of issues we have been examining have been puzzled over for decades in their modern form and for millennia in their ancient incarnations. Systematic analysis of the anthropic cosmological principle has, over time, revealed that the basic concept—that the universe is life-friendly—actually encompasses four separate but related subprinciples:

1. The "weak anthropic principle," which merely asserts in tautological fashion that the universe we inhabit must perforce be life-friendly since it happens to be inhabited by living observers like ourselves

2. The "strong anthropic principle," which states that the eventual emergence of life and intelligence in the universe is actually

predestined by the laws and constants of inanimate nature

3. The counterintuitive "participatory anthropic principle," which hypothesizes, on the strength of the Copenhagen interpretation of quantum mechanics, that observer-participancy is necessary to summon the universe into existence and to give it structure

4. The "final anthropic principle," which advances the extraordinary claim that once life has arisen anywhere in this or any other universe, its sophistication and pervasiveness will expand inexorably and exponentially until life's domain is coterminous with the boundaries of the cosmos itself

Let us now explore each of these versions of the cosmological anthropic principle in more detail.

The Weak Anthropic Principle

The weak anthropic principle is little more than a tautology—a recitation of a self-evident proposition. *Of course* the universe must be hospitable to life-forms like ourselves, this version holds, or else we humans would not be around to observe the universe. Carbon-based life-forms like human beings could no more live in a cosmos with five or six extended spatial dimensions (where stable planetary orbits would be impossible) or with extra-strength gravity than they could live on the surface of a star or in the vacuum of outer space. Why should we not logically expect that the cosmos we inhabit would be life-friendly? What other qualities could it possibly possess and still play host to our species or indeed to any carbon-based life-forms remotely like ourselves?

This argument becomes dicier when one contemplates the sheer improbability of the "just so" fine-tuning of the laws and constants of physics necessary to render them life-friendly. As Martin Rees put it in *Just Six Numbers*,

> A few basic physical laws set the "rules"; our emergence from a simple Big Bang was sensitive to six "cosmic numbers." Had these numbers not been "well tuned," the gradual unfolding of layer upon layer of complexity would have been quenched. Are there an infinity of other universes that are "badly tuned," and therefore sterile? Is our entire universe an "oasis" in a multiverse? Or should we seek other reasons for the providential value of our six numbers?[44]

How do proponents of the weak anthropic principle respond to the incontestable fact that even a tad bit of difference in the arbitrary constants of nature (like the strength of gravity or the value of the cosmological constant) would have doomed life to nonexistence and the universe to sterility? There are three typical responses.

The first is to simply ignore the congeries of coincidences that must conspire and collaborate in order to make the cosmos hospitable to life and to intone, mindlessly and relentlessly, "Of course the universe must be life-friendly or else there would be no life in it!" This amounts to nothing more than the utterance of a tautology.

The second response is to assert that the conspiracy of cosmic coincidences we witness is not so extraordinary after all because our universe may simply be one small island in a "multiverse," an inconceivably vast population of universes forever isolated from one another, each of which exhibits a different set of physical laws and constants and only a tiny minority of which could possibly harbor life. (The concept of a multiverse is common to various areas of physics. Under one version of cosmic inflation theory, for example, the vast ensemble of universes known as the multiverse is spawned by constant explosions of new Big Bangs in existing universes that create countless generations of baby universes. Under a very different theoretical approach—the so-called many worlds interpretation of quantum theory—the universe splits off into an infinite series of parallel universes each time a quantum observation occurs, thus populating an inconceivably numerous multiverse ensemble.)

The problem with this scenario is that it rests on the extravagant assumption that a multiverse actually exists—a hypothesis for which there is no empirical evidence whatsoever and which is, by definition, virtually impossible to either validate or disprove. Undaunted by the seemingly unscientific character of the multiverse theory (the hallmark of genuine science is, after all, that it offers predictions that can be either verified or rejected on the basis of empirical evidence), prominent defenders like Nobel laureate Steven Weinberg speculate that it may offer a kind of "random accident" explanation for the appearance of the fine-tuning of physical laws and thus may put meat on the embarrassingly bare bones of the weak anthropic principle:

> Anthropic arguments may become very important when applied to what we usually call the universe. Cosmologists increasingly speculate that just as the earth is just one of many planets, so also our big bang, the great expansion of the universe in which we live, may be just one of many big bangs that go off sporadically here and there in a much larger mega-universe. They further speculate that in these many different big bangs some of the supposed constants of nature take different values, and perhaps even some of what we now call the laws of nature take different forms. In this case, the question why the laws of nature that we discover and the constants of nature that we measure are what they are would have a rough teleological explanation—that it is only with this sort of big bang that there would be anyone to ask the question.[45]

The final response of defenders of the weak anthropic principle is to assert that some deep, still undiscovered physical principle lies beneath the apparent mystery of anthropic fine-tuning. Perhaps the laws and constants of our physics are the *only* ones that could, in principle, emerge from the Big Bang. Perhaps invariant laws of mathematics disallow any reshuffling of the fundamental parameters of nature, and the fact that those parameters also happen to be life-friendly is a meaningless epiphenomenon—inconsequential noise in the grand symphony of mathematically inevitable cosmic laws. It is the fond hope of a vast majority of scientists, including Weinberg, that this will turn out to be the case and that no deep connection will emerge between the shape of a so-called final theory of physical reality and the apparent fine-tuning of the laws and constants of the universe that render it oddly hospitable to carbon-based life.

The Strong Anthropic Principle

The strong version of the anthropic principle asserts that the laws of nature encode a hidden life-generating subtext and that the emergence of both life and intelligence from inanimate matter is thus inevitable, or at least highly likely. Few scientists currently subscribe to this quasi-religious notion, in part because it fails to specify a plausible natural mechanism and in part because it seems incapable of generating falsifiable predictions. As we shall see later, both defects can be remedied and a scientifically plausible statement of the strong anthropic principle can be formulated.

The Participatory Anthropic Principle

The participatory anthropic principle is the brainchild of a brilliant and unorthodox Princeton physicist, John Wheeler. Wheeler, who counts among his protégés the late Nobel laureate Richard Feynmann, has developed a version of the anthropic principle that takes to extraordinary lengths the Copenhagen interpretation of quantum measurement. To understand Wheeler's theory, a brief digression into the strange world of quantum theory is required.

One of the queerest aspects of quantum theory is that tiny subatomic particles like electrons are presumed to exist in a ghostlike state called a "superposition" until they are actually observed. Not until such particles are actually observed are they forced to "choose" their key characteristics. Under the Copenhagen interpretation (named for the birthplace of quantum physics), the very act of measurement and observation affects basic aspects of subatomic reality. No one truly understands how this could possibly be so, and alternative interpretations (some truly bizarre) have been advanced to explain the odd phenomenon of observer-participancy in the very structure of our physical reality. Einstein himself was convinced that quantum theory described only a puzzling artifact of a deeper classical reality yet to be uncovered and commented famously, "The theory yields a lot, but it hardly brings us any closer to the secrets of the Old One. In any case I am convinced that *He* does not throw dice."[46] The ongoing debate over the Copenhagen interpretation and its various alternatives is endlessly fascinating but need not detain us. For our purposes, it is enough to understand that the Copenhagen interpretation of quantum physics, at its core, implies that the act of observing reality actually influences the nature of reality.

John Wheeler has taken this core concept to its logical conclusion and articulated a startling vision of the distant future that has been dubbed the "participatory anthropic principle." It was summarized by Wheeler in the 1989 Santa Fe Institute lecture referred to previously ("It From Bit").[47] Wheeler had speculated in an earlier lecture that while "[t]he anthropic principle superficially looks like a tautology,"[48] it may in fact be subject to genuine prediction and thus "to destruction in the sense of Karl Popper."[49] (See sidebar, page 20.)

With this assumption in mind, Wheeler proceeded to consider a unique and utterly intriguing explanation of the "just so" coincidences that make carbon-based life and intelligence possible in our universe:

> Is the machinery of the universe so set up, and from the very beginning, that it is guaranteed to produce intelligent life at some long-distant point in its history-to-be? And is this proposition testable? . . . Perhaps. But how should such a fantastic correlation come about between big and small, between machinery and life, between future and past? . . . [H]ow can history ever have made things come out right, ever given a world of life, ever thrown up a communicating community of the kind required for the establishment of meaning? In brief, how can the machinery of the universe ever be imagined to get set up at the very beginning so as to produce man now? Impossible! Or impossible unless somehow—preposterous idea—meaning itself powers creation. But how? Is that what the quantum is all about?[51]

In "It From Bit," Wheeler expanded on this breathtaking speculation:

> *It from bit.* Otherwise put, every it—every particle, every field of force, even the spacetime continuum itself—derives its function, its meaning, its very existence entirely—even if in some contexts indirectly—from the apparatus-elicited answers to yes or no questions, binary choices, *bits.* It from bit symbolizes the idea that every item of the physical world has at bottom—at a very deep bottom, in most instances—an immaterial source and explanation; that what we call reality arises in the last analysis from the posing of yes-no questions, and the registering of equipment-evoked responses; in short, that all things physical are information-theoretic in origin and this is a *participatory universe.*[52]

Three elements of Wheeler's vision of a participatory universe are particularly relevant for our purposes. First is his concept of the cosmos as an autocatalytic loop. He states, "To endlessness [infinite regress] no alternative is evident but loop, such a loop as this: Physics gives rise to observer-participancy; observer-participancy given rise to information; and information gives rise to physics."[53] Second is his notion of the universe not as a machine, but as a self-organized system: "Directly opposed to the concept of universe as machine built on law is the vision of a world self-synthesized. On this view, the notes struck out on a piano by the observer-participants of all places and all times, bits though they are, in and by themselves constitute the great wide world of space and time and things."[54] Third, there is what Wheeler calls the "super-Copernican" principle:

> The super-Copernican principle . . . rejects now-centeredness in any account as firmly as Copernicus repudiated here-centeredness. It repudiates most of all any tacit adoption of here-centeredness in assessing observer-participants and their numbers. . . . We today, to be sure, through our registering devices, give a tangible meaning to the history

> of the photon that started on its way from a distant quasar long before there was any observer-participancy anywhere. However, the far more numerous establishers of meaning of time to come have a like inescapable part—by device-elicited questions and registration of answer—in generating the "reality" of today. For this purpose, moreover, there are billions of years yet to come, billions on billions of sites of observer-participancy yet to be occupied. How far foot and ferry have carried meaning-making communication in fifty thousand years gives faint feel for how far interstellar propagation is destined to carry it in fifty billion years.[55]

Taken together, these mind-boggling concepts comprise the participatory anthropic principle. The principle offers an explanation for the life-friendly qualities of our cosmos that is, at least superficially, diametrically opposed to the notion encapsulated in the strong anthropic principle (i.e., that the laws of nature were somehow fine-tuned from the outset to eventually yield life and intelligence). In Wheeler's remarkable vision, the device that fine-tunes the universe consists not of a precise initial blueprint, but of a vast assembly of billions upon billions of living observer-participants, the overwhelming majority of whom inhabit the distant future. It is the collective and retroactive effect of their countless acts of observation that reaches backward in time and creates our world, along with all of its physical laws and constants. Here is Wheeler's riveting description of the key difference between the two concepts:

> Surely—big bang and gravitational collapse advise us—the laws of physics cannot have existed from everlasting to everlasting. They must have come into being at the one gate in time, must fade away at the other. But at the beginning there were no gears and pinions, no corps of Swiss watchmakers to put things together, not even a preexisting plan. If this assessment is correct, every law of physics must be at bottom like the second law of thermodynamics, higgledy-piggledy in character, based on blind chance. Physics must be in the end law without law.[56]

Wheeler's startling scenario is generally disfavored by mainstream scientists. However, his key insight may turn out to be correct: that living and thinking creatures are, in at least some manner, vital participants in the inconceivably vast process of cosmic evolution.

The Final Anthropic Principle

The "final anthropic principle" is generally associated with the work of John Barrow and Frank Tipler. In the concluding chapter of their encyclopedic treatise *The Anthropic Cosmological Principle*, they set forth what they call the Omega Point theory of the final state of the universe. They begin their analysis by noting that while "there is no evidence whatsoever of intelligent life having any significant effect upon the Universe in the large," this may not always be the case:

> We *know* space travel is possible. We argued [previously] that even interstellar travel is possible. Thus once space travel begins, there are, in principle, no further physical barriers to prevent *Homo sapiens* (or our descendants) from eventually expanding to colonize a substantial portion, if not all, of the visible Cosmos. Once this has occurred, it becomes quite reasonable to speculate that the operations of all these intelligent beings could begin to affect the large scale evolution of the Universe. If this is true, it would be in this era–in the far future near the Final State of the Universe–that the true significance of life and intelligence would manifest itself. Present-day life would then have cosmic significance because of what future life may someday accomplish.[57]

Barrow and Tipler proceed to speculate that in the far-distant future, the boundary of the biosphere will expand to be coterminous with that of the cosmos itself:

> Finally, the time is reached when life has encompassed the entire Universe and regulated all matter contained therein. Life begins to manipulate the dynamical evolution of the universe as a whole, forcing the horizons to disappear, first in one direction, and then another.

> The information stored continues to increase. . . . From our [prior] discussion . . . we see that if life evolves in all of the many universes in a quantum cosmology, and if life continues to exist in all of these universes, then *all* of these universes, which include *all* possible histories among them, will approach the Omega Point. At the instant the Omega Point is reached, life will have gained control of all matter and forces not only in a single universe, but in all universes whose existence is logically possible; life will have spread into *all* spatial regions in all universes which could logically exist, and will have stored an infinite amount of information, including *all* bits of knowledge which it is logically possible to know. And this is the end.[58]

The final anthropic principle postulates that this final glorious state of the cosmos—the Omega Point—in which the brute physical reality of a sterile universe is entirely subdued and transcended by the self-organizing powers of life and intelligence is somehow implicit in the laws of physics from the very beginning. In this respect, the final anthropic principle bears a close kinship to the strong anthropic principle.

So there we have it—the principal permutations of the cosmological anthropic principle. It must be conceded that all the various anthropic ways of thinking are disfavored by most mainstream physicists. (One scientist waggishly disparaged the whole enterprise of anthropic reasoning as "CRAP—completely ridiculous anthropic principles.") These traditionalists contend that even the tamest version of the principle—the weak anthropic principle—is nothing but a lazy man's excuse to stop searching for a final theory of everything that would embed the seemingly arbitrary physical constants of nature in a mathematically unique and logically inevitable system of invariant rules. The dismissive views of the late physicist Heinz Pagels are typical of those of many traditionalists:

> Physicists and cosmologists who appeal to anthropic reasoning seem to me to be gratuitously abandoning the successful program of conventional physical science of understanding the quantitative properties

> of our universe on the basis of universal physical laws. Perhaps their exasperation and frustration . . . has gotten the better of them. The influence of the anthropic principle on the development of contemporary cosmological models has been sterile: it has explained nothing; and it has even had a negative influence, as evidenced by the fact that the values of certain constants . . . for which anthropic reasoning was once invoked as an explanation can now be explained by new physical laws. . . . I would opt for rejecting the anthropic principle as needless clutter in the conceptual repertoire of science.[59]

For the traditionalists, the deep mystery of the life-friendly qualities of the cosmos is a scientifically irrelevant and intellectually dangerous distraction—or at the very most, an indication that we have not searched hard enough or long enough to uncover the final mathematical secrets of the universe.

Chapter Three

Hints of a Fourth Law

The Second Law of Thermodynamics

The Second Law of Thermodynamics states that entropy in a closed system can never decrease. Entropy is commonly defined as the opposite of the quantity of energy available to perform work—that is, as the quantitative measure of the amount of thermal energy not available to perform work in a closed system.

It is often contended that life violates the Second Law of Thermodynamics because there appears to be a decrease in the entropy of a living system. However, this is not true. Life merely takes advantage of energy gradients in nonequilibrium open systems to power its self-organization and replication. The source of

(continued on next page)

If a system of deep-seated principles underlies the mysterious capacity of the cosmos to self-organize and achieve ever greater levels of complexity, then that system may differ profoundly from the set of physical laws with which scientists are currently familiar. One scientist who has dared to think far outside the box and ponder how and why the cosmos seems hell-bent on overcoming the disintegrative force of entropy is Stuart Kauffman of the Santa Fe Institute.

A physician by training, a polymath by inclination, and now a highly successful business entrepreneur by choice, Kauffman is searching for what he calls the "origins of order" in both the earthly biosphere and in the universe at large. The origins of order, Kauffman believes, may consist of a "fourth law of thermodynamics for self-constructing open thermodynamic systems such as a biosphere."[1] Further, Kauffman thinks that this fourth law may capture a fundamental and hitherto unrecognized force in the universe that relentlessly drives the cosmos as a whole to construct itself to be as complex and diverse as possible. This suspected fourth law would serve as a counterforce to the disorder-generating Second Law of Thermodynamics, which seems to drive all closed systems to a state of featureless equilibrium and maximum entropy. (See sidebar, page 49.)

Kauffman lives and works in the spiritually rich landscape of Santa Fe, New Mexico. He is one of the founders of the study of complexity and is a perfect embodiment of the strengths as well as the weaknesses of this new field of inquiry. Loquacious, captivating, and often downright off-the-wall, Kauffman fearlessly rushes in where traditional scientists fear to tread. Among colleagues, Kauffman is not only admired for his courage to tackle issues that seem beyond the pale of traditional science but also occasionally derided for his irrepressible instinct for intellectual adventure. In *The End of Science* journalist John Horgan likened Kauffman's style of speaking about scientific issues to "the improvisations of a jazz musician" in that it is "short on melody, long on digression."[3]

Moreover, Kauffman is not shy about turning his expertise in the sciences of complexity into a source of personal wealth. A company he cofounded with a prominent accounting firm employs complexity science-based computer simulations and other tools to help Fortune 500 companies solve business problems that seem intractably complicated.

Kauffman's first love, however, is not making money but making sense of the cosmos. He was first drawn to his quest to uncover a fourth law of thermodynamics by a central mystery in evolutionary theory. That mystery is nicely captured by the phrase of a prominent biologist of the early twentieth century, Hugo De Vries, who proclaimed that what really interested him was not the *survival* of the fittest but the *arrival* of the fittest. In other words, how did the beautifully intricate and interdependent ballet of molecules through which DNA is replicated and proteins are assembled first get composed? How did the universal coding scheme of DNA, common to virtually all living creatures, arise in the first place? How did Mother Nature first assemble the lavish palette of options from which the grim reaper of natural selection could prune away all but the choicest samples?

In probing the mystery of the *arrival* of the fittest, should we be searching for rules of self-organization that precede the operation of the great Darwinian principle of natural selection? Is this self-ordering process altogether different from the process of random mutation, which is the only source of novelty condoned by orthodox evolutionists? And, most intriguingly, is a force of self-organization the silent but indispensable partner of Darwin's force of natural selection, performing the role of nature's creative experimentalist by offering up a dizzying array of diverse living forms and processes upon which Darwinian evolution may operate?

This question was the focus of Kauffman's first book, *The Origins*

the energy gradient can be sunlight (in the case of surface organisms) or chemical reactions (in the case of organisms inhabiting subsurface habitats like those adjoining deep-ocean thermal vents).

Beginning in the mid–nineteenth century, the dismal implications of the Second Law for the ultimate fate of the universe began to be understood. The German physicist Hermann von Helmholtz, was the first to suggest the horrid prospect of an ultimate "heat death" of the universe. The heat death would occur in the distant future when all thermodynamic processes have exhausted themselves and the cosmos settles into a bland, featureless uniformity. It is a measure of the power of abstract cosmological concepts to affect (or at least resonate with) popular culture that the notion of an eventual cosmic heat death led to statements of despair on the part of popular theologians as well as to calls for true believers to reject any form of science/religion accommodation.

of Order. In it he sought nothing less than to fashion an entirely new tapestry of fundamental biological theory, weaving the impressive capacity of complex adaptive systems to spontaneously self-organize into the stiff fabric of orthodox Darwinian theory:

> What are the sources of the overwhelming and beautiful order which graces the living world? . . . One view, Darwin's, captivates us all: natural selection and the great branching tree of life, spreading from the major phyla to the minor genera and species, to terminal twigs, to curious humans seeking their place. Darwin and evolutionism stand astride us, whatever the mutterings of creation scientists. But is the view right? Better, is it adequate? It is not that Darwin is wrong but that he got hold of only part of the truth. For Darwin's answer to the sources of the order we see all around us is overwhelmingly an appeal to a singular force: natural selection. It is this single-force view which I believe to be inadequate, for it fails to notice, fails to stress, fails to incorporate the possibility that simple and complex systems exhibit order spontaneously. . . . What is mysterious is the extent of such spontaneous order in life and how such self-ordering may mingle with Darwin's mechanism of evolution—natural selection—to permit or, better, to produce what we see.[4]

To understand how the inherent tendency of certain natural processes to achieve order might relate to the traditional concept of natural selection as the engine of evolution, Kauffman concluded, three key questions must be answered:

- What are the spontaneous sources of order?
- How do the forces driving systems to spontaneous order enable and collaborate with the forces of natural selection?
- What properties of complex living systems confer upon them the capacity to adapt and evolve?

Modern statements of the Second Law (and of entropy) are couched in terms of statistics. Under these formulations, even the most unlikely far-from-equilibrium states can appear and reappear in a closed system over absurdly long periods of time. Given an infinite expanse of time, it is certain that every such state will recur an infinite number of times. The implications of this puzzling phenomenon for the anthropic cosmological principle are just now beginning to be explored by physicists like Leonard Susskind of Stanford University, who wonders whether it may give credence to the notion that an "unknown agent intervened in the evolution [of the Universe] for reasons of its own."[2]

Do We Need a Fourth Law to Explain a Non-ergodic Cosmos?

Before deciding that we must contemplate a

(continued on next page)

Kauffman's analysis focuses in large part on phenomena that were the precursors to biochemistry: autocatalytic webs of complex carbon-based polymers. One of his key goals is to approach the study of the mystery of life's origin by analyzing the inherent self-ordering propensity of such polymer sets:

> [W]e do not now know how life may have started. Any discussion is at best a body of ideas. The central problem is this: How hard is it to obtain a self-reproducing system of complex organic molecules, capable of a metabolism coordinating the flow of small molecules and energy needed for reproduction and capable of further evolution? Contrary to all our expectations, the answer, I think, is that it may be surprisingly *easy*. To state it another way, I want to suggest that we can think of the origin of life as an expected *emergent collective property* of a modestly complex mixture of catalytic polymers, such as proteins or catalytic RNA, which catalyze one another's formation. I believe that the origin of life was not an enormously improbable event, but law-like and governed by new principles of self-organization in complex webs of catalysts.[5]

Perhaps the most important point to emerge from Kauffman's analysis is that the propensity for self-organization he has identified appears to be a basic attribute of matter itself, provided that matter is arrayed in sufficiently complex arrangements and that those complex arrays have sufficient opportunity to interact with one another. A second major point developed by Kaufmann is that this mysterious force of self-organization, which appears to be a generic property of complex adaptive systems perched at the "edge of chaos" (i.e., halfway between stasis and completely random disorder), may help explain the speed with which evolution operates to elevate organisms and entire ecosystems to improbable levels of complexity and order:

> To some great extent, evolution is a complex combinatorial optimization process in each of the coevolving species in a linked ecosystem, where the landscape of each actor deforms as the other actors move.

possible Fourth Law of Thermodynamics to explain the nonergodic, far-from-equilibrium current state of the cosmos, it is important to remember that the entropic principle codified in the Second Law of Thermodynamics applies only if the system under consideration is closed. If the system is open (like the far-from-equilibrium ecosystems of Earth that are energized by sunlight and chemical energy), then the Second Law is not contradicted by phenomena evidencing an increase in order and organization and a local decrease in entropy.

The system that is our cosmos may possibly be "open" within the meaning of the Second Law because the energy unleashed by the Big Bang may be the cosmological equivalent of life-powering sunlight on Earth—a source of disequilibrium so potent that it continues to fuel the expansion and diversification of the universe to this day, some 15 billion years after the initial cataclysm. Another possibility, favored by MIT cosmologist Alan

Guth, is that the vast quantity of positive energy unleashed by the Big Bang is precisely counterbalanced by negative energy represented by the gravitational field of the cosmos. The tension between these two great energy fields could conceivably serve as the central clock spring that keeps the universe running, without any violation of the Second Law.

The fascinating question then becomes this: How did the universe ever get wound up in the first place into the mother of all nonergodic conditions—the vastly far-from-equilibrium state represented by the moment of the Big Bang itself?

Within each organism, conflicting constraints yield a rugged fitness landscape graced with many peaks, ridges, and valleys. Two major alternative limitations to selection exist. First, selection is limited by the structure of the fitness landscape it acts on; in many landscapes, as the organisms under selection grow more complex, the attainable optima fall below the average features of the class of systems on which selection is acting. Second, on any landscape, a mutation-selection balance is struck; beyond some level of complexity, selection cannot hold an adapting population at the high peaks of the landscape, and the population then falls toward the average properties of the underlying class of systems. Both limitations suggest that, in sufficiently complex systems, much of the order found is that spontaneously present in the class of systems under selection. Therefore, I have made bold to suggest that much of the order seen in organisms is precisely the spontaneous order in the systems of which we are composed. Such order has beauty and elegance, casting an image of permanence and underlying law over biology. Evolution is not just "chance caught on the wing." It is not just a tinkering of the ad hoc, of bricolage, of contraption. It is emergent order honored and honed by selection.[6]

This exciting new perspective suggests several possibilities:

- The capacity to evolve may itself be capable of evolving, as Oxford biologist Richard Dawkins persuasively argued in a 1988 paper entitled "The Evolution of Evolvability,"[7] eventually yielding the phenomenon of cultural evolution.

- The primary focus of study should be *coevolution* of organisms because the most important components of a fitness landscape are likely not to be inanimate features but coevolving creatures and plants that constitute the ecosystem that constrains and shapes evolution.

The Central Dogma of Molecular Biology

In a frequently quoted passage, Nobel laureate Francis Crick, codiscoverer of the molecular structure of DNA, memorialized what he called

(continued on next page)

- Coevolution offers unique opportunities to ratchet up the complexity of organisms and entire ecosystems because a complicated system affording temporary evolutionary success (say, echolocation in bats) tends to elicit complicated countermeasures, which in turn elicit yet more complex patterns of predation.

- The fitness landscape itself coevolves along with the creatures within it, in part because it is itself partially alive (insofar as it consists of a menagerie of coevolving organisms) but also because life has the capacity to reshape geological features and indeed the inanimate content of the atmosphere and the ocean.

- If the tape of life and evolution were rerun—either on Earth or on another planet—the same impulse toward order and self-organization would emerge and the fundamental pattern of ever greater biological complexity would eventually reveal itself.

What Kauffman believes he has begun to uncover is a kind of fundamental grammar of a self-assembling cosmos—a universal "law of construction" that dictates a pattern of increasing complexity—rooted in self-ordering principles that constitute a deep-seated counterweight to the mandate of ever increasing entropy enshrined in the Second Law of Thermodynamics. It is thus hardly surprising that Kauffman has sought to extend his daring new paradigm, quite literally, to the ends of the universe.

In two subsequent books—*At Home in the Universe*[8] and *Investigations*[9]—Kauffman ponders the deeper implications of the perplexing tendency of nature to produce what he calls "order for free." Kauffman prefaces his grander themes with a familiar overture: that the self-ordering tendency of nature forces us to take seriously the possibility of a biological paradigm sharply at odds with the evolution-as-tinkerer scenario favored by traditional Darwinists:

the central dogma of molecular biology: "The Central Dogma . . . states that once information has passed into protein it cannot get out again. In more detail, the transfer of information from nucleic acid to nucleic acid, or from nucleic acid to protein may be possible, but transfer from protein to protein, or from protein to nucleic acid is imposible."[17]

The central dogma undergirds classical Darwinian theory as well as contemporary restatements of it by Richard Dawkins and others. It sharply distinguishes evolutionary theory from Lamarckism (the notion that acquired characteristics can be inherited). It is an article of faith among adherents to the central dogma that, in Dawkins's words, "the Lamarckian theory is completely wrong."[18]

> The existence of spontaneous order is a stunning challenge to our settled ideas in biology since Darwin. Most biologists have believed for over a century that selection is the sole source of order in biology, that selection alone is the "tinkerer" that crafts the forms. But if the forms selection chooses among were generated by laws of complexity, then selection has always had a handmaiden. It is not, after all, the sole source of order, and organisms are not just tinkered-together contraptions, but expressions of deeper natural laws. If all this is true, what a revision of the Darwinian worldview will lie before us! Not we the accidental but we the expected.[10]

Such deeper natural laws, Kauffman suspects, must somehow relate to the tendency of the universe, in the billions of years since the Big Bang, to achieve not bland uniformity, but a state of profound nonequilibrium:

> I am a doctor-biologist. I am hopeful enough to think I might possibly help in understanding the origin of life and its subsequent evolution. I am not a physicist. I am not brazen enough to presume to think about cosmic evolution. But I wonder: Whence cometh all this bubbling activity and complexity? Ultimately, it must be a natural expression of a universe that is not in equilibrium, where instead of the featureless homogeneity of a vessel of gas molecules, there are differences, potentials, that drive the formation of complexity. The flash of the Big Bang 15 billion years ago has yielded a universe said to be expanding, perhaps never to fall together in the Big Crunch. It is a *nonequilibrium* universe filled with too many hydrogen and helium atoms compared with the most stable atomic form, iron. It is a universe of galaxies and clusters of galaxies on many scales, where none might have formed at all. It is a universe of stunningly abundant free energy available for performing work.[11]

And what does this nonequilibrium state of the universe tell us about the origin of life and the central mechanism driving the phenomenon of evolution? Perhaps a great deal, Kauffman suspects:

> The life around us must somehow be the natural consequence of the coupling of that free energy to forms of matter. How? No one knows. But we shall hazard hypotheses along the way. Here is no mere scientific search. Here is a mystical longing, a sacred core first sought around that small campfire sometime in the past 3 million years. This way lies the search for our roots. If we are, in ways we do not yet see, natural expressions of matter and energy coupled together in nonequilibrium systems, if life in its abundance were bound to arise, not as an incalculably improbable accident, but as an expected fulfillment of the natural order, then we truly are at home in the universe.[12]

This is heady stuff but it is by no means the apogee of Kauffman's speculative extravagance. In *Investigations*, he suggests that he has glimpsed hints of "something like a fourth law, a tendency for self-constructing biospheres to enlarge their workspace, the dimensionality of their adjacent possible, perhaps as fast, on average, as is possible."[13] The suspected "fourth law" of thermodynamics would be a mysterious mirror version of the familiar Second Law of Thermodynamics, which states that the entropy of a closed system either is constant or increases. Thus, while the Second Law reflects the simple statistical principle that isolated physical systems will tend to flow from less probable to more probable states, the conjectured fourth law seeks to explain the incontestable fact that the universe at present is in what physicists call a "nonergodic" condition—a vastly nonuniform, nonequilibrium state characterized by *both* white-hot stars and parsecs of empty space, by dead clouds of interstellar gas and at least one teeming planetary biosphere. (See sidebar, page 51.)

Kauffman concludes by elevating his fourth law to a possible fundamental principle of the physical universe:

> The universe, in short, is breaking symmetries all the time by generating such novelties, creating distinctive molecules or other forms which had never existed before. Indeed, there may be a general law for bio-

> spheres and perhaps even the universe as a whole along the following lines. A candidate fourth law: As an average trend, biospheres and the universe create novelty and diversity as fast as they can manage to do so without destroying the accumulated propagating organization that is the basis and nexus from which further novelty is discovered and incorporated into the propagating organization.[14]

A clear implication of this scenario, as Kauffman acknowledges, is that the emergence of life itself is prefigured in the fundamental physical laws of nature, that "pregnant in the birth of the universe was the birth of life."[15]

After considering Kauffman's "proto-science" (a term he uses to confess both humility and hubris), it is interesting to revisit the evolutionary thinking of Charles Darwin. It is by now nearly forgotten (or conveniently overlooked) that Darwin himself did not view the force of natural selection, indisputably powerful as it might be, as fully sufficient to generate the near miraculous novelty and diversity of the living world. Darwin, like Kauffman, always assumed that natural selection had a handmaiden, a silent partner, a generative order-building force that operated in tandem with the destructive force that weeded out less fit individuals and species.

For Darwin, this second force was the currently discredited idea of Lamarckian evolution: the notion that use or disuse of particular bodily parts could influence the shape and character of such parts in an organism's offspring. Indeed, Darwin vehemently objected to what he viewed as a serious misrepresentation of his evolutionary theory as resting solely on the principle of natural selection:

> I have now recapitulated the facts and considerations which have thoroughly convinced me that species have been modified, during a long course of descent. This has been effected chiefly through the natural selection of numerous successive, slight, favourable variations; aided in an important manner by the inherited effects of the use and disuse of parts; and in an unimportant manner, that is in relation to adaptive

> structures, whether past or present, by the direct action of external conditions, and by variations which seem to us in our ignorance to arise spontaneously. It appears that I formerly underrated the frequency and value of these latter forms of variation, as leading to permanent modifications of structure independently of natural selection. But as my conclusions have lately been much misrepresented, and it has been stated that I attribute the modification of species exclusively to natural selection, I may be permitted to remark that in the first edition of this work, and subsequently, I placed in a most conspicuous position—namely, at the close of the Introduction—the following words: "I am convinced that natural selection has been the main but not the exclusive means of modification." This has been of no avail. Great is the power of steady misrepresentation; but the history of science shows that fortunately this power does not long endure.[16]

While the second evolutionary mechanism favored by Darwin—Lamarckian revision of the genome—may be disallowed by what Nobelist Francis Crick has called the "central dogma of molecular biology" (the notion, now disputed at least with respect to bacteria, that biological information travels on a one-way street from genome to phenotype and cannot travel back "upstream" from phenotype to genome), the key point is that Darwin, like Kauffman, believed that natural selection, by itself, was insufficient to generate the near miraculous diversity of the biosphere that surrounds and sustains us. (See sidebar, page 53.) It is a point almost universally ignored by the so-called ultra-Darwinists of the present era. But one can surely hope that the musings of Kauffman and other evolutionary contrarians will eventually overcome the "power of steady misrepresentation" of Darwin's theory that Charles Darwin himself so deeply deplored.

The contours of Kauffman's theories are themselves the evolutionary product of a very special cultural environment he inhabits and within which his novel ideas have been nourished: the Santa Fe Institute, unofficial global headquarters for the study of the sciences of complexity. The unique culture of the institute, which is steeped in a vision of biology and evolution as universal explanatory paradigms, tends to focus the attention of its disciples intently on a conception of nature as an interactive set of complex adaptive systems that exhibit the near miraculous capacity to generate novel emergent phenomena (like life, consciousness, and civilization) at ascending levels of organization of matter and energy. Leading acolytes of the institute like Stuart Kauffman are thus inherently predisposed toward a *biologically* inspired vision of the ultimate nature of reality—a vision which we shall explore in greater detail in the next chapter.

The Coming Fusion of Biology and Cosmology

Part 1 of this book prompts the perplexing question of how our universe could possibly have emerged from the cataclysm of the Big Bang in such an improbably biofriendly condition. Was the life-sustaining quality of our universe merely a random accident—an extraordinarily lucky roll of the dice in an endless cosmic crap game that cosmologists call eternal chaotic inflation—or was it the consequence of deeply moving causal factors we do not yet understand?

In search of a scientific answer to the mystery posed by the very existence of an anthropic universe, part 2 of the book examines critically the main ideas that biologists, physicists, and other scientists have put forward in recent years by way of explanation.

BIOCOSM

Chapter Four

The Dreams of Christian de Duve

The Templeton Prize

The Templeton Prize for Progress Toward Research or Discoveries about Spiritual Realities is awarded annually to recognize progress in religion. The prize, whose amount of £700,000 sterling is deliberately set above the sum awarded Nobel laureates, aims at rewarding scientists and theologians who have, in the view of Templeton Foundation prize judges, achieved "breakthroughs in religious concepts and knowledge." In the words of the Templeton Foundation, the purpose of the prize is this:

How might humankind's spiritual information and advancement increase by more than 100 fold? This is the challenge presented by the Templeton Prize. Just as

(continued on next page)

Each historical epoch is associated with a prevailing cosmological metaphor—a canonical paradigm that embodies a collective societal vision of the ultimate nature of the universe. As one prominent scientist has noted, these metaphors invariably reflect the supreme technological or cultural accomplishments of a particular era. For example, for the ancient Greeks, the preeminent technology was musical-instrument construction and Greek cosmology was dominated by the concept of harmony, order, and the music of the spheres. For that supremely gifted playwright of the Elizabethan era, William Shakespeare, the world was "a stage, and all the men and women merely players."

In Isaac Newton's era, the highest technological achievement was clockwork manufacture and, unsurprisingly, the Newtonian metaphor for the universe was a great cosmic timepiece, ticking away with inexorable precision throughout eternal time and infinite space. In a subsequent era, the thermal power of internal and external combustion was harnessed to fuel the Industrial Revolution and the dominant cosmological metaphor became a vision of the cosmos as a vast thermodynamic engine facing the inevitable fate of heat death.

During this same era, when Darwin's great theory commenced its steady rise, the metaphor of nature as an unforgiving chess master was popularized by evolution advocate Thomas Henry Huxley: "The chess board is the world, the pieces are the phenomena of the universe, the rules of the game are what we call the laws of Nature. The player on the other side is hidden from us. We know that his play is always fair, just, and patient. But also we know, to our cost, that he never overlooks a mistake, or makes the smallest allowance for ignorance."[1]

"Today," astrophysicist and Templeton Prize–winner Paul Davies has observed, "the computer is the pinnacle of technology, so now it's fashionable to talk about nature as a computational process."[2] (See sidebar, page 63.) This metaphor portrays the movements of every material object from quark to galaxy as subroutines in a vast computing system running on software that consists of the fundamental laws and constants of nature.

What will tomorrow's dominant cosmological metaphor be? If, as genetics pioneer Craig Ventner forecasts, the twenty-first century will be celebrated by future historians as the "century of biology,"[4] could it be that the biological sciences—including the biologically inspired sciences of complexity—will supply the prevailing cosmological paradigm in the decades ahead? Some leading thinkers believe that this is a distinct possibility. If so, Stuart Kauffman's biologically inspired speculations about the nature of the universe may turn out to be a prescient prologue to the emerging cosmological models of the twenty-first century.

One of the most creative contemporary thinkers seeking to catalyze a creative fusion of biology and cosmology is biochemist and Nobel laureate Christian de Duve. In *Vital Dust*, published in 1994, de Duve issued a daunting challenge to biologists and cosmologists, urging them to unify their seemingly incommensurable intellectual realms:

> Traditionally, the dialogue with philosophers has been held mainly by theoretical physicists and mathematicians, probably because of a common meeting ground in abstraction. The resulting cosmological picture comprised all facets of the physical world, from elementary particles to galaxies, but either ignored life or had life and mind tagged on to the picture as separate entities by some implicit, sometimes explicit, recourse to vitalism and dualism. This is wrong. Life is an integral part of the universe; it is even the most complex and significant part of the known universe. The manifestations of life should dominate our world picture, not be excluded from it. This has become particularly mandatory in view of the revolutionary advances in our understanding of life's fundamental processes.[5]

knowledge in science, medicine, cosmology and other disciplines has grown exponentially during the past century, the Templeton Prize honors and encourages the many entrepreneurs trying various ways for discoveries and breakthroughs to expand human perceptions of divinity and to help in the acceleration of divine creativity. Their various methods, particularly through scientific research, serve to supplement the wonderful ancient scriptures and traditions of all the world's religions. Many honors and titles and prizes have been given for many centuries and will be given in the future for good works, reconciliation, saintliness or for relief of poverty and sickness. But these very worthy endeavors are not the purpose of the Templeton Prize. Instead, this award is intended to encourage the concepts that resources and manpower are needed to accelerate progress in spiritual discoveries, which can help humans to learn over 100 fold more about divinity. We hope that by learning about the lives of the awardees, millions of people will be uplifted and inspired toward research and more discoveries about aspects of divinity. The Prize is intended to help people see the infinity of the Universal Spirit still creating the galaxies and all living things and the variety of ways in which the Creator is revealing himself to different people. We hope all religions may become more dynamic and inspirational.[3]

Recent recipients of the prize include physicists Freeman Dyson and Paul Davies as well as science-and-religion synthesizers like Arthur Peacocke and Ian Barbour.

The essence of de Duve's vision is that inanimate matter is indelibly imprinted with a life-giving template, that every atom of cosmic dust is vitally infused with a life-generating force. De Duve's optimistic scenario contrasts starkly with that of his fellow European, the French scientist Jacques Monod (like de Duve, a Nobel laureate), who sprinkled his ostentatiously bleak pièce de résistance—a pretentious but beautifully written book entitled *Chance and Necessity*—with exclamations of existential pathos worthy of Albert Camus or Jean-Paul Sartre. The universe, Monod wrote, "was not pregnant with life, nor the biosphere with man."[6] Indeed, mankind has been revealed to be utterly alone in the universe, bereft even of the comforting illusion of a benevolent creator that sustained humanity's cultural childhood: "The ancient covenant is in pieces: man knows at last that he is alone in the universe's unfeeling immensity, out of which he emerged only by chance. His destiny is nowhere spelled out; nor is his duty. The kingdom above or the darkness below: it is for him to choose."[7]

De Duve's response to Monod's pessimism is simple and stark:

> My reasons for seeing the universe as meaningful lie in what I perceive as its built-in necessities. Monod stressed the improbability of life and mind and the preponderant role of chance in their emergence, hence the lack of design in the universe, hence its absurdity and pointlessness. My reading of the same facts is different. It gives chance the same role, but acting within a stringent set of constraints as to produce life and mind obligatorily, not once but many times. To Monod's famous sentence, "The universe was not pregnant with life, nor the biosphere with man," I reply: "You are wrong. They were."[8]

For de Duve, the genesis of life was no accidental pregnancy but a result foreordained by the most basic laws of nature. "From the perspective of determinism and constrained contingency that pervades the history of life as I have reconstructed it," de Duve concludes, "life and mind emerge not as the results of freakish accidents, but as natural manifestations of matter, written into the fabric of the universe."[9]

It is not only rudimentary life but conscious thought as well that is, in de Duve's view, an inevitable consequence of cosmic processes: "Conscious thought belongs to the cosmological picture, not as some freak epiphenomenon peculiar to our own biosphere, but as a fundamental manifestation of matter. Thought is generated and supported by life, which is itself generated and supported by the rest of the cosmos."[10] Indeed, in an eerie foreshadowing of the hypothesis to be advanced in this book, de Duve went so far as to conclude that the concept of life and the idea of the cosmos were inseparably linked, at least in his own mind: "The universe *is* life, with the necessary infrastructure around; it consists foremost of trillions of biospheres generated and sustained by the rest of the universe."[11]

In the years that have elapsed since the publication of de Duve's courageous call for cross-disciplinary dialogue among biologists, cosmologists, and philosophers, the first hints of a revolutionary new biology-centered cosmological model have begun to emerge. This new paradigm, foreshadowed in earlier speculations by cosmologist Stephen Hawking, physicist John Wheeler, quantum theorist Erwin Schrödinger, Princeton astrophysicist Freeman Dyson, and cosmologist Fred Hoyle, rests on a sober assessment of the astonishing array of "just so" coincidences inherent in the physical characteristics of our universe —characteristics which render the cosmos peculiarly friendly to carbon-based life. Freeman Dyson offered this prediction in 1985: "I am suggesting that there may come a time when physics will be willing to learn from biology as biology has been willing to learn from physics, a time when physics will accept the endless diversity of nature as one of its central themes, just as biology has accepted the unity of the genetic coding apparatus as one of its central dogmas."[12]

The time foreseen by Dyson may be rapidly approaching. With the advent of the new biologically focused scientific paradigm of self-organizing complexity and with the publication of groundbreaking cosmological speculations like cosmologist Lee Smolin's *The Life of the Cosmos*,[13] a serious effort to undertake a grand unification of biology

Darwin and Artificial Selection

Darwin's Origin of Species *is centered on the notion that natural selection is the primary, though not exclusive, engine driving speciation and evolution in the wild. Yet what is often forgotten is that "artificial selection" furnished the key metaphor that allowed Darwin to fashion his explanation of the phenomenon of the appearance of a succession of genetically related species in the geological record.*

Artificial selection, in Darwin's view, encompassed two distinct practices: deliberate selective breeding of animal and plant species to achieve desired characteristics and "a form of Selection, which may be called Unconscious, and which results from every one trying to possess and breed from the very best individual animals."[21] This unconscious process of artificial selection is the "more important"[22] of the two types because it requires no conscious thought or intention. As

Darwin put it, "Thus, a man who intends keeping pointers naturally tries to get as good dogs as he can, and afterwards breeds from his own best dogs, but he has no wish or expectation of permanently altering the breed."[23]

This process of artificial selection frequently renders unrecognizable the root species from which a vast range of domesticated varieties have been derived. States Darwin, "A large amount of change, thus slowly and unconsciously accumulated, explains, as I believe, the well-known fact, that in a number of cases we cannot recognize, and therefore do not know, the wild parent-stocks of the plants which have been longest cultivated in our flower and kitchen gardens."[24]

Darwin's consummate achievement was to realize that nature could likewise engage in the practice of unconscious selection in a manner analogous to that of human breeders and domesticators:

(continued on next page)

and cosmology appears to be commencing. Even skeptics like physicist and Nobel Prize winner Steven Weinberg—who concluded his popular account of the "first three minutes" with the extravagantly gloomy observation that "[t]he more the universe seems comprehensible, the more it also seems pointless"[14]—have been swept along by the new intellectual tide. As reported in the January 1999 issue of *Scientific American*:

> One direction, explored recently by Steven Weinberg of the University of Texas at Austin and his colleagues, invokes the last resort of cosmologists, the anthropic principle. If the observed universe is merely one of an infinity of disconnected universes—each of which might have slightly different constants of nature, as suggested by some incarnations of inflationary theory combined with emerging ideas of quantum gravity—then physicists can hope to estimate the magnitude of the cosmological constant by asking in which universes intelligent life is likely to evolve.[15]

More recently, Cambridge physicist Neil Turok has conducted an exhaustive analysis of cosmic inflationary theory, which seeks to explain the flatness and homogeneity of the observable universe as the consequence of an extremely rapid expansion of space shortly after the Big Bang. His startling conclusion: some version of the anthropic principle is required in order to furnish a satisfactory explanation of the peculiarly life-friendly qualities of the observed universe. Turok, who collaborates with Stephen Hawking to develop the world's most sophisticated mathematical models of the Big Bang, concedes that it would "be nice if the correct theory predicted only Universes very like ours" but has concluded that this "may be too much to ask."[16] In short, Turok suggests, "we may be forced to pursue the more limited goal of a theory which allows of all types [of universes]" and explains the unusual character of the universe we actually inhabit as dictated by the requirement that it must be user-friendly to observers like ourselves![17] In other words, the anthropic principle may turn out to be the foundational paradigm underlying a future final theory of physics and cosmology!

Unfortunately, the anthropic principle, at least as currently formulated, raises far more questions than it answers. For instance, one common but unsatisfactory response to the disconcerting necessity of admitting there are indeed anthropic constraints on solutions showing a wide range of mathematically possible laws and constants of nature, only a tiny subset of which are life-friendly, is to superimpose a cosmological theory known as "eternal chaotic inflation" on the weak variant of the cosmological anthropic principle. This approach, identified with Steven Weinberg, asserts that "the expanding cloud of billions of galaxies that we call the big bang may be just one fragment of a much larger universe in which big bangs go off all the time, each one with different values for the fundamental constants."[18] It is no more a mystery that our particular branch of the cosmos exhibits life-friendly physical characteristics, according to Weinberg, than that life evolved on the hospitable Earth "rather than some horrid place, like Mercury or Pluto."[19]

There are two difficulties with this approach. First, universes spawned by Big Bangs other than our own are inaccessible from our own universe, at least with the experimental techniques currently available to scientists. Second, the Weinberg approach simply takes refuge in a brute, unfathomable mystery—the conjectured lucky roll of the dice in a crap game of eternal chaotic inflation—and declines to probe seriously into the possibility of a naturalistic cosmic evolutionary process that has the capacity to yield a life-friendly set of physical laws and constants on a nonrandom basis. It is rather as if Charles Darwin, contemplating the fabled tangled bank, had confessed not a magnificent obsession with gaining an understanding of the mysterious natural processes that had yielded "endless forms most beautiful and most wonderful,"[20] but rather a smug satisfaction that *of course* the earthly biosphere must have somehow evolved in a just-so manner mysteriously friendly to human life, or else Darwin and other humans would not be around to contemplate it.

Indeed, the situation that confronts cosmologists today is eerily reminiscent of that which faced biologists before Charles Darwin propounded his revolutionary theory of evolution. Darwin confronted the

Variability is not actually caused by man; he only unintentionally exposes organic beings to new conditions of life, and then nature acts on the organization and causes it to vary. But man can and does select the variations given to him by nature, and thus accumulates them in any desired manner. He thus adapts animals and plants for his own benefit or pleasure. He may do this methodically, or he may do it unconsciously by preserving the individuals most useful or pleasing to him without any intention of altering the breed. It is certain that he can largely influence the character of a breed by selecting, in each successive generation, individual differences so slight as to be inappreciable except by an educated eye. This unconscious process of selection has been the great agency in the formation of the most distinct and useful domestic breeds. That many breeds produced by man have to a large extent the character of natural species, is shown by the inextricable doubts whether many of them are varieties or aboriginally distinct species. There is no reason why the principles which have acted so efficiently under domestication should not have acted under nature. In the survival of favored individuals and races, during the constantly-recurrent Struggle for Existence, we see a powerful and ever-acting form of Selection.[25]

seemingly miraculous phenomenon of a "fine-tuned" natural order in which every creature and plant appeared to occupy a unique and well-designed niche. Refusing to surrender to the brute mystery posed by the appearance of nature's design, Darwin masterfully deployed the art of metaphor to elucidate a radical hypothesis—the origin of species through natural selection—that explained the apparent miracle as a natural phenomenon. (See sidebar, page 66.)

The metaphor furnished by the familiar process of *artificial* selection was Darwin's crucial stepping stone. Indeed, the practice of artificial selection through plant and animal breeding was the primary intellectual model that guided Darwin in his quest to solve the mystery of the origin of species and to demonstrate in principle the plausibility of his theory that variation and natural selection were the prime movers responsible for the phenomenon of the speciation. (It is a nice irony that Darwin, condemned by his detractors as the destroyer of poetry and mystery in nature, employed the poetic device of metaphor to elucidate one of the most important scientific advances of all time!)

So, too, today a few venturesome cosmologists have begun to use the same poetic tool utilized by Darwin—the art of metaphorical thinking—to develop novel intellectual models that might offer a logical explanation for what appears to be an unfathomable mystery: the apparent fine-tuning of the cosmos. If one of these scientists should succeed in this long-odds quest, he or she may well earn a rank in the pantheon of science equal to that enjoyed by Darwin.

The cosmological metaphor chosen by these iconoclastic theorists is life itself. What if life, they ask in the spirit of de Duve, were not a cosmic accident but the essential reality at the very heart of the elegant machinery of the universe? What if Darwin's principle of natural selection were merely a tiny fractal embodiment of a universal life-giving principle that drives the evolution of stars, galaxies, and the cosmos itself? What if the universe were literally in the process of coming to life?

At the heart of this radical new paradigm is the concept of "emergence"—a key element of the new sciences of complexity pioneered by

Consilience

What if the diverse and insular realms of human knowledge—the natural sciences, "soft" disciplines like sociology and anthropology, even the humanities, the arts, ethics, and religion—could be melded seamlessly into a unitary whole? What if the vast store of insight and wisdom accumulated by our species over the millennia could be fused indissolubly into a single

(continued on next page)

the Santa Fe Institute. Emergence has been described as follows by complexity theorist John Holland: "The hallmark of emergence is this sense of much coming from little. . . . We are everywhere confronted with emergence in complex adaptive systems—ant colonies, networks of neurons, the immune system, the Internet, and the global economy, to name a few—where the behavior of the whole is much more complex than the behavior of the parts."[26]

Four generic features of the phenomenon of emergence are noteworthy in the context of the quest for a satisfactory explanation of an anthropically fine-tuned universe. First, as Holland notes, "[t]he possibilities for emergence are compounded when the elements of the system include some capacity, however elementary, for adaptation or learning."[27] Second, the "component mechanisms [in an emergent system] interact without central control."[28] Third, the "possibilities for emergence increase rapidly as the flexibility of the interactions increases."[29] Fourth, and perhaps most important, "persistent patterns at one level of observation can become building blocks at still more complex levels,"[30] yielding a defining characteristic of emergent systems as embodying "*hierarchical organization* (configurations of generators become generators at a higher level of organization)."[31]

The fourth characteristic of emergent systems is especially crucial because it implies that the number of hierarchical levels underlying a particular emergent phenomenon can be indefinitely large and that sufficiently complex multilevel hierarchies of basic components as simple as quarks and subatomic particles (the initial products of the Big Bang) are, in proper combination, capable of eventually yielding such high-level phenomena as human culture (including specialized domains of that culture like scientific inquiry). Says Holland,

> [H]uman creative activity, ranging from the construction of metaphors through innovations in business and government to the creation of new scientific theories, seems to involve a controlled invocation of emergence.[32]

towering pyramid of intellectual and cultural achievement?

At the base would surely be the physical sciences and mathematics, proud exemplars of precision and transparency. The next tier in the ascending pyramid would be occupied by the biological disciplines, gaining rapidly in rigor and predictive power as the result of dramatic research initiatives like the Human Genome Project and the new computer-aided field of artificial life. One level up on the pyramid of human achievement would reside the social sciences—economics, psychology, and related fields—blending subtly into evolutionary theory in hybrid disciplines like sociobiology. Still higher on the knowledge pyramid one would find the arts and the humanities, followed at the apex by law, ethics, and religion.

That, in a nutshell, is the Enlightenment dream of "consilience"—the holistic notion that all branches of human knowledge are fundamentally linked and constitute a deeply coherent, self-reinforcing intellectual system. The search for consilience within the sciences and across the uneasy frontier that demarcates the great science-humanities continental divide has been the lifelong quest of Harvard biologist Edward O.Wilson. Admirers and detractors of Wilson's monumental Sociobiology *(a treatise that applied Darwinian principles to the analysis of human culture) will find in Wilson's vision of consilience the distilled essence of all they admire or disdain, as the case may be, in the philosophy of this towering and controversial thinker.*

Here are some of the large issues Wilson seeks to resolve with the aid of this vision: Can ethical imperatives be derived from the findings of evolutionary biology? Can social scientists apply the principles of Darwinian struggle to comprehend the birth and death of the world's great religions? Are the arts reducible to the unfolding of inborn rules

(continued on next page)

The same point about the indefinitely large hierarchical layering potential of emergent complex adaptive systems has been made implicitly by Nobel laureate and Santa Fe Institute scientist Murray Gell-Mann:

> Examples on Earth of the operation of complex adaptive systems include biological evolution, learning and thinking in animals (including people), the functioning of the immune system in mammals and other vertebrates, the operation of the human scientific enterprise, and the behavior of computers that are built or programmed to evolve strategies—for example by means of neural nets or genetic algorithms. Clearly, complex adaptive systems have a tendency to give rise to other complex adaptive systems.[33]

There is no indication that the tendency noted by Gell-Mann has been arrested at this particular historical moment. On the contrary, the evidence is overwhelming that the process of multilevel hierarchical emergence is accelerating rapidly.

How far can the process of emergence propel the phenomenon of technological complexification in theory? How high can this process allow mankind and its progeny to "climb Mount Improbable" (in Richard Dawkins's felicitous phrase),[34] employing only the ropes and pitons furnished by the principles of complexity theory and evolution? Several recent speculations about the potential magnitude of the ongoing process of cosmological emergence indicate that it may, in principle, be capable of extending the reach of life to the very limits of the physical universe. In other words, at some point in the far distant future (called Point Omega by Barrow and Tipler) the domain of living matter and the domain of the physical cosmos itself may become coterminous.

If this scenario turns out to be plausible, it implies that one cannot offer credible predictions about the physical state of the universe in the distant future without taking into account the possible physical impact of a cosmologically extended biosphere on the cosmos as a whole. Much as living processes have shaped the atmosphere and geochemistry of the Earth, life and intelligence in the future may profoundly alter the very shape and character of the universe itself.

This scenario implies as well that the nature and characteristics of living matter, including preeminently life's defining qualities of counterentropic organization, replication, and evolution, may be linked with the ultimate fate of the cosmos. As Freeman Dyson put it in an essay published in the March 28, 2002, issue of the *New York Review of Books*, "Could it happen that life will grow in scope and power until it dominates the ecology of the universe as it today dominates the ecology of planet Earth? Could it happen that life will dominate the universe to such a degree as to achieve mastery over the geometry of space and time and create a habitat in which life can survive forever?"[35]

To seek to explain the potential unity of life and the physical cosmos in the far future is to commence a search for a new style of final theory. If ever fully articulated, such a theory would differ dramatically from the traditional conception of an ultimate "theory of everything" in which the laws and constants of nature are uniquely specified by invariant and self-evident principles of mathematics. A theory in harmony with the dreams of Christian de Duve would, like Darwin's masterwork, take seriously the profound impact of the biosphere over immense spans of time on the patterns of organization of physical matter, as well as the unpredictable impact of chance and fortuity in shaping such patterns. In this respect, it would be a historical theory in precisely the same sense that the theory of evolution is historical. Such a theory would be consonant with de Duve's view of inanimate nature as "vital dust" and would wholeheartedly embrace the radical thought eloquently voiced by theoretical chemist Michael Polanyi that "this universe is still dead, but it already has the capacity of coming to life."[36]

Such a theory would unify conceptually the realms of life and nonlife and then go on to assert forcefully the preeminence of the former. Such a theory would triumphantly vindicate Harvard biologist Edward O. Wilson's vision of consilience: a conjunction of all the sciences in a seamless web of causal structure and interlinked inference.[37] (See sidebar, page 69.) And finally, such a theory would add to the list of ontological

of mental development? And, most fundamentally, did human culture and the human genome coevolve across the vast expanse of humanity's "deep history"—those darkest of ages when our species made the epic journey from ape to man—in such a way as to fundamentally shape our spirit as well as our physiognomy?

The objective student of the history of science will be forced to concede, I think, that Wilson is onto something profound. The natural, biological, and social sciences are indeed rapidly integrating across disciplinary lines. Influenced by novel scientific paradigms emerging from the new metadiscipline of complexity theory and empowered by potent computer modeling tools, many social scientists and biologists are working to impart a hitherto inconceivable degree of rigor to "soft" disciplines like sociology and anthropology.

possibilities traditionally contemplated by cosmologists—that the universe may be a great equation, a great computation, or a great accident—the astonishing possibility that the cosmos may be quintessentially a vast unfolding life.

Chapter Five

Biology and Teleology: Deciphering the Utility Function of the Cosmos

Deciphering God's Utility Function

In an intriguing chapter in his 1995 book River Out of Eden, *ultra-Darwinist Richard Dawkins offers his speculations about "God's utility function." By this tantalizing phrase, Dawkins means the objective that nature appears to be attempting to maximize. Here is how Dawkins explains his audacious quest:*

"Utility function" is a technical term not of engineers but of economists. It means "that which is maximized." Economic planners and social engineers are rather like architects and real engineers in that they strive to maximize something. . . . It isn't always obvious what

(continued on next page)

The notion of the universe as a living creature is not new. Nor is the concept that nature as a whole shares with living organisms the quality of teleological progression—an idea, associated primarily with the philosophy of Aristotle, that the evolutionary development of a system is drawn irresistibly toward a predetermined purpose or goal.

The canonical view of the contemporary science establishment is, in the words of evolutionary biologist Ernst Mayr, that "Darwin's theory of natural selection made any invocation of teleology unnecessary."[1] The Darwinian paradigm, Mayr has opined, utterly subverts the notion that seemingly end-directed processes in living creatures—let alone similar processes in the realm of inanimate matter—can be profitably investigated from a teleological perspective: "Processes in living organisms owe their apparent goal-directedness to the operation of an inborn genetic or acquired program. Adapted systems, such as the heart or kidneys, may engage in activities that can be considered goal seeking, but the systems themselves were acquired during evolution and are continuously fine-tuned by natural selection."[2]

Indeed, according to Mayr, the Darwinian perspective extinguishes the possibility of a teleological explanation of the behavior of nature at a cosmic level: "Finally, there was a belief [before Darwin] in cosmic teleology, with a purpose and predetermined goal ascribed to everything in nature. Modern science, however, is unable to substantiate the existence of any such cosmic teleology."[3] Perhaps, but the question remains: Has modern science ruled out such a possibility?

While Mayr congratulates Darwin for rendering this possibility literally unthinkable by any scientifically literate person, one wonders whether Mayr has departed the venue of hypothesis-driven science, where the concept of truth is inherently provisional, and entered the realm of ideology, where truth possesses a quality of finality and certainty bordering on religious faith. He declares, "Cosmic teleology, an intrinsic process leading life automatically to ever greater perfection, is fallacious, with all seemingly teleological processes explicable by purely material processes; and determinism is thus repudiated, which places our fate squarely in our own evolved hands."[4] The ideological quality of Mayr's thinking is starkly revealed by assertions like the following concerning the allegedly sharp boundary demarcating the realms of life and nonlife:

> A most important principle of the new biological philosophy . . . is the dual nature of biological processes. These activities are governed both by the universal laws of physics and chemistry and by a genetic program, itself the result of natural selection, which has molded the genotype for millions of generations. The causal factor of the possession of a genetic program is unique to living organisms, and is totally absent in the inanimate world.[5]

How can Mayr possibly know, with absolute certainty, that this assertion is unassailably true? How can Mayr dismiss out of hand the possibility, lyrically expressed by the late physicist Heinz Pagels, that the appearance of life and intelligence arising in the universe is essentially foreordained by the flawlessly coherent details of what Pagels called the "cosmic code"? And how can Mayr and his ilk possibly overlook the evidence—or, at the very least, the appearance—of directionality in evolution: the sense that the force of evolution propels life inexorably toward ever-greater complexity, diversity, mastery over its environment, and, eventually, consciousness? It is intellectually perverse to insist that the record of life on Earth reveals indisputable proof of a lack of teleology and that Darwin's great theory will admit of no other interpretation.

While Mayr's views capture nicely the distilled essence of orthodox Darwinism, there are reputable thinkers who disagree utterly with his stark conclusions regarding the obsolescence of teleological explana-

individuals, or firms, or governments are striving to maximize. But it is probably safe to assume that they are maximizing something. This is because Homo sapiens is a deeply purpose-ridden species. The principle holds good even if the utility function turns out to be a weighted sum or some other complicated function of many inputs. Let us return to living bodies and try to extract their utility function. There could be many but, revealingly, it will eventually turn out that they all reduce to one. A good way to dramatize our task is to imagine that living creatures were made by a Divine Engineer and try to work out, by reverse engineering, what the Engineer was trying to maximize: What was God's Utility Function?[7]

For Dawkins, the unassailable truth is that "the true utility function of life, that which is being maximized in the natural world, is DNA survival."[8] A natural world dominated by this utility function displays, in Dawkins's views, a pitiless indifference to human life and human values:

If the universe were just electrons and selfish genes, meaningless tragedies . . . are exactly what we should expect, along with equally meaningless good fortune. Such a universe would be neither good nor evil in intention. It would manifest no intentions of any kind. In a universe of blind physical forces and genetic replication, some people are going to get hurt, other people are going to get lucky, and you won't find any rhyme or reason in it, nor any justice. The universe we observe has precisely the properties we should expect if there is, at bottom, no design, no purpose, no evil and no good, nothing but blind, pitiless indifference.[9]

Dawkins never explains why he assumes that DNA survival—a utility function restricted, as far as we know, to the earthly biosphere—should be uncritically

(continued on next page)

tion. The science essayist James Gleick, for instance, offers the suggestion that Darwin's approach is unique in science precisely *because* it embraces a teleological frame of reference. While the triumph of the scientific approach in physics and astronomy is a story of the *displacement* of teleological explanation by reductionist physical causes, the Darwinian scenario is distinctive, according to Gleick, precisely *because* it looks to the fitness of the *final product* of the interplay of such forces as the basis for explaining the phenomenon of evolution:

> In science, on the whole, physical cause dominates. Indeed, as astronomy and physics emerged from the shadow of religion, no small part of the pain came from discarding arguments by design, forward-looking teleology—the earth is what it is so that humanity can do what it does. In biology, however, Darwin firmly established teleology as the central mode of thinking about cause. The biological world may not fulfill God's design, but it fulfills a design shaped by natural selection. Natural selection operates not on genes or embryos, but on the final product. So an adaptationist explanation for the shape of an organism or the function of an organ looks to its *cause,* not its physical cause but its final cause. Final cause survives in science wherever Darwinian thinking has become habitual. A modern anthropologist speculating about cannibalism or ritual sacrifice tends, rightly or wrongly, to ask only what purpose it serves.[6]

Indeed, the ultra-skeptical French biologist and Nobel Prize winner Jacques Monod, hardly a softheaded romanticist when it came to discerning the distinguishing features of living creatures, concluded that a key defining characteristic of life was that it exhibited the property of "teleonomy," that is, that the DNA molecules that comprise an organism's genome prescribe the organism's future project or purpose—its future ontogenetic pathway and the form of its mature phenotype.

Let us concede, then, that there is at least a possibility that biology in general and evolutionary analysis in particular may be profitably investigated from a teleological perspective. Such a perspective prompts us to ask this simple question: What is the "utility function" of the process under investigation? By utility function we mean the outcome that is maximized by that process. (See sidebar, page 75.)

For example, consider Richard Dawkins's notion of the "selfish gene."[10] An ultra-Darwinist if there ever was one, Dawkins nonetheless offers an inherently teleological perspective when he suggests that the utility function of natural selection is the replication and survival of "selfish genes." The differential survival and persistence of information encoded in such genes is the so-called blind-watchmaker mechanism that sculpts the bodies, the extended phenotypes, and the genetically prescribed behaviors of every living organism—past, present or future.[11] It is not the mere biochemistry of interacting genes and proteins that offers the most satisfying explanation of the evolutionary process but rather the *survival advantage* conferred by variations, small or large, in the immensely complex biochemical symphony by which an organism develops from a DNA recipe to a mature, reproducing adult.

In a similar vein, MIT cognitive scientist Steven Pinker observed in his 2002 book *The Blank Slate* that the very existence of a kind of teleological causation in the evolutionary process helps explain the existence of seemingly antisocial behaviors and predispositions like a male propensity to sexual promiscuity and unhealthy gluttony. He stated, "The difference between the mechanisms that impel organisms to behave in real time and the mechanisms that shaped the design of the organism over evolutionary time is important enough to merit some jargon."[12]

In a remarkable passage with which the great teleologist Aristotle would have felt quite comfortable, Pinker described the key difference between these two sorts of causal mechanisms: "A *proximate* cause of behavior is the mechanism that pushes behavior buttons in real time, such as the hunger and lust that impel people to eat and have sex. An *ultimate* cause is the adaptive rationale that led the proximate cause to evolve, such as the need for nutrition and reproduction that gave us the drives of hunger and lust."[13] The operation of evolution cannot be coherently understood, Pinker maintains, absent a clear appreciation of the differences between these two categories of causation:

equated to God's cosmic utility function.

Attractors and the Cosmic Code

A technical definition of attractor was provided by complexity theorist Stuart Kauffman in The Origins of Order: *"The general definition of an attractor is a set of points or states in state space to which trajectories within some volume of state space converge asymptotically over time."*[15] *An attractor can be visualized as the bottom of a "basin of attraction"—think of the drain at the bottom of a sink toward which the water in the sink flows because of the attractive force of gravity.*

Chaos and complexity theorists use the concept of attractors to formalize a wide range of phenomena, including prebiotic evolution of complex polymers and the near miraculous process of biological ontogeny (which is the developmental history of an individual organism, both embryonic and postnatal). In ontogenesis, the key attractor is the mature organism whose emergence culminates the ontogenetic process.

By analogy, a cosmic attractor would be an end state or fundamental cyclic process toward which intermediate cosmic processes tend to converge. The basic laws of physics—the cosmic code—would presumably drive such convergence just as the DNA instruction set—the genetic code—drives the process of ontogenesis toward the form of a mature organism.

> The distinction between proximate and ultimate causation is indispensable in understanding ourselves because it determines the answer to every question of the form "Why did that person act as he did?" To take a simple example, ultimately people crave sex in order to reproduce (because the ultimate cause of sex is reproduction), but proximately they may do everything they can not to reproduce (because the proximate cause of sex is pleasure).[14]

As biology evolves into a more physically and mathematically sophisticated discipline, potentially able to contribute ever more valuable insights and methodological perspectives to the "hard" sciences like physics and biochemistry, the continuing dominance of the Darwinian paradigm in contemporary biological science may augur a rebirth of the teleological perspective in the physical sciences—in particular, in cosmology. If so, we may profitably begin to ask not merely what are the proximate physical causes of cosmological phenomena like the Big Bang and the seemingly fine-tuned constants of nature, but also what is their conceivable utility function and even their ultimate cause?

If Christian de Duve's conception of inanimate matter as "vital dust" is a harbinger of the ascendance of biology-inspired cosmological models, then it seems at least plausible that the ancient and seemingly discredited Aristotelian notion of teleological causation could soon enjoy a renaissance in the field of cosmology. From the teleological perspective, the ultimate nature of cosmological reality may be best captured by hypotheses concerning the possible final outcomes of cosmological processes. Under this paradigm, those outcomes might be viewed as what complexity theorists call attractors—patterns toward which ostensibly undirected physical processes converge robustly. (See sidebar, page 78.)

Like the ontogeny of a living creature, where the genetic code specifies a unique recipe for constructing a complex mature organism, the evolution of the cosmos might then be usefully viewed as a teleological process, incorporating a discernible utility function and guided by the celestial counterpart of earthly DNA (i.e., Heinz Pagels's cosmic code). The cosmos would then hypothetically exhibit the property that Monod ascribed exclusively to living creatures—teleonomy—with the future purpose or project of the cosmos inscribed indelibly in the basic laws of physics and constants of nature that comprise the cosmic code. Deciphering the utility function of that cosmic code could then become a key goal of the science of cosmology.

Chapter Six

Professor Smolin's Incredible Reproducing Universe

The Principle of Mediocrity and the Phenomenon of Rarity

The principle of mediocrity is a statistically based rule of thumb that, absent contrary evidence, a particular sample (our planet, for instance, or our universe) should be assumed to be a typical example of the ensemble of which it is a part. Some elements of an ensemble are demonstrably atypical, however. A particular planet may be far more predisposed to generate carbon-based life than most others. In a 2000 book entitled Rare Earth: Why Complex Life Is Uncommon in the Universe,[4] *two University of Washington scientists—paleontologist Peter T. Ward and*

(continued on next page)

A few years ago the astrophysicist Lee Smolin asked me why I had such a passionate interest in both cosmology (the study of the origin of the universe and its evolution) and the emerging role of regions in geopolitics.[1] (In addition to my scientific pursuits and law practice, I serve as president of a nonprofit organization called the Conference of World Regions, which is attempting to develop an understanding of both the constructive and destructive role likely to be played by autonomy-minded regions like the Basque Country, Catalonia, and Bavaria in the emerging world order of the twenty-first century.)

I told Professor Smolin that the unifying factor linking these seemingly disparate passions was that both topics—the role of regions in the global economy and the profound mysteries surrounding the coevolution of congeries of celestial bodies in the cosmic void—presented an opportunity to use complexity theory to try to understand the development of self-organizing adaptive systems. Both global regions and phenomena of interest to cosmologists, I told Smolin, constituted complex adaptive systems in which the seemingly unguided interactions of autonomous agents often resulted in startling phenomena of "emergence": the appearance of novel patterns of hierarchical organization whose characteristics could not have been predicted in advance of their appearance (at least not by any human mind).

Professor Smolin did not seem inclined to dismiss my explanation as zany (or worse), perhaps because he himself had already drawn a striking analogy between the cosmos and that quintessential artifact of human culture, a great city. In the concluding pages of his masterwork, *The Life of the Cosmos*, Smolin likened the universe to a vast unplanned metropolis—endlessly creative, miraculously self-catalyzing, forever tantalizing in its mysterious capacity to achieve novelty and complexity: "The metaphor of the universe we are trying now to imagine, which I would like to set against the picture of the universe as a clock, is an image of the universe as a city, as an endless negotiation, an endless construction of the new out of the old. No one made the city, there is no city-maker, as there is a clockmaker. If a city can make itself, without a maker, why can the same not be true of the universe?"[2]

For Smolin, as for Darwin, the art of metaphor serves as an essential tool to be used in the fashioning of a novel scientific conjecture. The two key metaphors that illuminate Darwin's *Origin of Species*—Malthusian competition between human beings scrapping for scarce resources as a metaphor for the battle for survival of species in the wild and artificial selection as a metaphor for natural selection—were at the heart of his attempt to depict the origin of new species as the logical consequence of *natural* selection. As Harvard evolutionary theorist Ernst Mayr has pointed out, the key contribution that British economist Thomas R. Malthus made to Darwin's thinking was the notion of the potentially exponential growth of populations of reproducing organisms, checked only by such harshly limiting factors as starvation, disease, and death: "The world of Malthus was a pessimistic world: there are ever-repeated catastrophes, an unending, fierce struggle for existence. . . . However much Darwin might have begun to question the benign nature of the struggle for existence, he clearly did not appreciate the fierceness of this struggle before reading Malthus."[3]

The principal metaphor that underlies Smolin's theory is captured perfectly in the title of his book: the *life* of the cosmos. By this title he emphatically does not mean to imply that the universe is actually alive but rather that it shares one of life's most important characteristics: differential reproductive success of offspring, yielding a phenomenon Smolin calls "cosmological natural selection."

What Lee Smolin has described is dramatic indeed: an evolutionary process operating at the largest conceivable scale. Under Smolin's theory, new baby universes are born in the hearts of black holes, which

astronomer Donald Brownlee—drew on fresh insights from the new interdisciplinary field of astrobiology to put forward a dramatic hypothesis: While life at the level of bacteria is likely to be ubiquitous throughout the cosmos, complex multicellular life (i.e., anything that crawls, walks, or flies) is probably vanishingly rare and perhaps even unique to Earth.

The key set of inferences that underlie what they call the "Rare Earth" hypothesis are these:

- *Our solar system is located in just the right "habitable zone" in the Milky Way galaxy, and our planet occupies an analogous sweet spot in the local solar neighborhood.*
- *Our unusually large moon, exceedingly rare for an Earth-like planet, played a crucial role in stabilizing the tilt of Earth's spin axis, which in turn stabilized seasonal changes.*
- *Perhaps most important, the sudden and startling*

proliferation of multicellular diversity 550 million years ago known as the Cambrian Explosion was sensitively dependent on two fortuitous circumstances, each of which required billions of years of preparation: the construction of an oxygen atmosphere and a large number of preceding evolutionary adaptations "to allow the evolution of an ocean liner—our animals—from the toy sailboat—the bacteria—that began it all."[5]

As Ward and Brownlee point out, leading figures in the scientific establishment are strongly biased toward the opposing view that not only complex but intelligent extraterrestrial life is pervasive throughout the universe—a corollary of the principle of mediocrity. Happily, this disagreement between the Rare Earth iconoclasts like Ward and Brownlee and the mediocrity mainstreamers is a

(continued on next page)

yield new Big Bangs on the "other side" of a black hole's singularity. Further, the constants and laws of physics vary slightly from baby universe to baby universe, yielding a natural selection process that favors the reproduction of universes adept at creating black holes and thus new baby universes. The birth of our own universe in a Big Bang approximately 15 billion years ago was, in Smolin's view, merely one instance of the ongoing process of cosmic reproduction.

This is a thrilling conjecture because it offers, at long last, a possible path forward toward an understanding of the central mystery: Why are the laws and constants of nature seemingly fine-turned to favor the appearance of carbon-based life? For Smolin the answer is that the appearance of a life-friendly universe is merely an epiphenomenon—a secondary and entirely coincidental consequence of a black hole–friendly universe. An evolutionary process that yields a maximum population of black holes per baby universe will also just happen to yield a maximum population of life-friendly baby universes, according to Smolin. (The explanation for the coincidence is complicated, but it centers on the propensity of black hole–friendly universes to produce copious quantities of carbon, which is obviously a prerequisite for the appearance of carbon-based life-forms like ourselves.)

Professor Smolin's courageous hypothesis has drawn predictable sniping from traditionalists in the astrophysics community. (At one gathering, he was asked facetiously whether sexual selection played any role in cosmological natural selection. He replied with a straight face that as far as he could tell, black holes did not need to engage in sex in order to give birth to baby universes!)

Because his ideas are so brave and iconoclastic, Professor Smolin's defenders (myself included) have hoped fervently that his conjecture might be able to provide a complete and satisfactory explanation for the life-friendly appearance of our cosmos. For if Smolin's ideas could be proven to be correct, then our cosmos could be seen as merely a statistically predictable product of a physically comprehensible evolutionary process. Far from being unlikely in the extreme, the configuration of laws and constants that we observe would be at the midpoint of a normal bell curve. The blind watchmaker of evolution alone would have fine-tuned the cosmological environment, just as the processes hypothesized by Darwin fine-tuned the earthly environment. Our cosmos could then be shown to be entirely ordinary, and there would be a wonderful consilience to our conception of nature at two very different levels of emergence. As an added bonus, what scientists call the "principle of mediocrity" would be triumphantly vindicated and the various versions of the cosmological anthropic principle could be finally consigned to the trash bin of intellectual history. (The principle of mediocrity, which is one of the most cherished legacies of the Copernican revolution, counsels that we should assume, absent contrary evidence, that the universe we inhabit is broadly representative of most others in a multiverse ensemble— that is, that it is a "mediocre" and thus a typical sample. See sidebar, page 81.)

Unfortunately, Smolin's grand theory suffers from two disastrous flaws. First, as British astronomer Martin Rees and other experts have pointed out, the physical laws and constants that prevail in our particular universe do not, in fact, appear to be fine-tuned to maximize black hole production.[7] Second, and more important, Smolin's hypothesis lacks any proposed mechanism of heredity that would ensure high-fidelity replication in baby universes of the cosmic code that prevails in the "mother" cosmos. Absent such a mechanism—equivalent to DNA replication in biology—a process of cosmic replication and natural selection is logically impossible. Why? Because as the pioneering computer genius John von Neumann demonstrated in a famous 1948 California Institute of Technology lecture entitled "On the General and Logical Theory of Automata," *any* self-reproducing object—human, mouse, or baby universe—must contain four fundamental components:

1. A *blueprint*, providing the plan for construction of offspring

2. A *factory*, to carry out the construction

3. A *controller*, to ensure that the factory follows the plan

dispute that could conceivably be resolved through experimentation and observation. Officials at the SETI Institute put it this way in a thoughtful commentary about the Rare Earth hypothesis: "While no experiment can prove we're alone, SETI experiments could show that we're not. Consequently, arguments about whether intelligence is a rare event in the universe—while instructive and interesting—may be akin to discussing with Columbus whether there's any chance his voyage will uncover a new continent. Discussion does not settle such arguments. Only experimentation can."[6]

4. A *duplicating machine*, to transmit a copy of the blueprint to the offspring[8]

In the context of Smolin's hypothesis, one can surmise that the physical laws and constants of our universe and its presumed progeny could conceivably constitute a von Neumann blueprint (literally a cosmic code, in Heinz Pagels's phrase) and the universe at large could serve as a sort of von Neumann factory. But what devices or natural processes could conceivably play the roles of von Neumann's controller and his duplicating machine in the context of a hypothesized process of cosmological replication and natural selection? These elements are needed to endow the hypothesized process of cosmological natural selection with what Richard Dawkins called in an exchange with Smolin a "strong phenomenon of heredity."[9]

Does this mean that Professor Smolin's idea of cosmological natural selection should be completely discarded? Absolutely not. What Smolin has proposed is an intriguing mechanism for importing the paradigm of biological natural selection into the realm of cosmology. His basic notion is that the birth and evolution of the cosmos is best understood by viewing it through the prism of a hypothesized process of selective cosmic replication. What is genuinely novel about Smolin's perspective is that it permits serious consideration of the possibility that this hypothesized process may have a *discernible utility function*, analogous to "selfish gene" survival and replication on Earth. While critics have credibly questioned whether the *particular* utility function favored by Smolin—black hole production—is in fact optimized by the laws and constants of nature that prevail in our universe, they have not thereby undermined the essence of his new paradigm: his vision of the universe as a self-organizing replicator that optimizes its reproductive success. It is this feature of Smolin's theory—the key idea that the phenomenon of cosmological replication and evolution may have a discernible utility function—that starkly distinguishes his ideas from those of astrophysicists like Stanford scientist Andrei Linde, who favor a process of "eternal chaotic inflation." (Linde's theory—which is essentially that new Big

Bangs are going all the time off beyond the visible limits of our universe and will continue doing so for all of eternity—is discussed in chapter 13.)

Can the essential structure of Smolin's idea of cosmological natural selection be salvaged, despite the evident flaws of the particular version of the idea he has espoused? That depends in large part on whether credible cosmological versions of the two logically necessary elements of any self-replicating von Neumann automaton—a controller and a duplicating device—can be formulated in the context of Smolin's general theory.

Could the missing von Neumann elements be secreted in the rich unfolding miracle of the evolving biosphere? Put differently, could life itself be far more than a mere metaphor for the evolving reality of the cosmos? Could, in fact, the origin and evolution of life and the emergence of intelligence in an inanimate universe be at the very heart of the mysterious process of cosmological ontogeny and replication hypothesized by Professor Smolin?

Part Three

Design, Complexity, and Evolution: An Eternal Cosmic Waltz

Part 2 of this book began to frame some of the key elements of a new vision of the cosmos. Of particular importance were, first, the notion that biological principles and patterns of terrestrial evolution might be telling us something important about the fundamental nature of the universe and, second, the idea that the we might actually be capable of discerning the utility function of the cosmos (i.e., the value that the immensely long process of cosmic evolution appears to be attempting to maximize).

Part 3 of the book builds on this analysis. It begins with an attempt to deal forthrightly with sensitive and sometimes inflammatory topics at the frontier of science and religion. After undertaking a risky journey into disputed territory—the cultural hot zone that is the situs of contemporary intellectual combat between evolutionists and creationists—the book goes on to propose a possible synthesis of these conflicting viewpoints. The proffered synthesis rests on the possibility that cosmic design could conceivably be coaxed from the primordial chaos by evolutionary forces operating over billions of years and at a cosmic scale.

BIOCOSM

Chapter Seven

Creationists vs. Evolutionists: Dispatches from a Hot Cultural War

Michael Behe and Irreducible Complexity

When intelligent design advocates are asked to supply the name of a recognized scientist who subscribes to their point of view, they inevitably mention Michael Behe. Behe is a religiously inclined biochemist who has argued strenuously that minute subsystems within living cells—bacterial flagella, for instance—are irreducibly complex because each of the elements of such subsystems are functionally dependent on other elements. As Behe puts it, "By irreducibly complex I mean a single system composed of several well-matched, interacting parts that contribute to the basic function, wherein the removal of any one of the parts

(continued on next page)

To pose deep questions about the relationship of life, intelligence, and the evolution of the universe is to risk the hazard of suffering serious collateral damage in the hot cultural war between contemporary creationists (who have taken to calling themselves "intelligent design [ID] theorists") and evolutionists. On both sides of the front line in this torrid conflict, ideologues attack one another relentlessly with what can only be described as religious fervor and undeviating faith in the absolute truth of their basic assumptions. In a bellicose spirit reminiscent of the intractable Arab-Israeli conflict, combatants on both sides of the epic evolution/creation struggle lob verbal grenades at one another across a sprawling battlefield that ranges from the Kansas Board of Education to the halls of Congress.

Questioning the proposition that blind natural selection is the sole and exhaustive explanation for the seeming miracle of evolution opens Pandora's box to all manner of cultish fantasies, the ultra-Darwinists warn. The great watchmaker that is biological evolution must be assumed to be utterly blind, they strenuously contend, or the magnificent edifice of Darwinian thought and theory will surely crumble.

The proponents of intelligent design—call them "IDers"—counter with the plea that biological systems are irreducibly complex, that living creatures exhibit a quality they call "specified complexity" (rather like an encrypted message or an intercepted communication from ET), and that the implacably atheistic bias of traditional Darwinism is robbing our culture of its capacity to inculcate a sense of the potential nobility of mankind. (See sidebar, page 89.) Most important, they contend, Darwinism, through its dogmatic devotion to materialism, threatens to extinguish our society's capacity to surrender to religious faith.

Scott versus Dembski

Both camps boast articulate defenders. The plucky Eugenie Scott, who heads that redoubt of Darwinian orthodoxy known as the National Center for Science Education, is often matched in debate against William Dembski, a seriously gifted mathematician and philosopher who is a Senior Fellow at Seattle's Discovery Institute, the nation's premier enclave of ID thinking. Following is a sample of the overheated rhetoric from both sides of the conflict.

The real objective of the IDers, claims Ms. Scott, is to infiltrate the biology classrooms of public schools and to clandestinely infect tender teenage minds with the heretical notion that a supernatural intelligence is the "intelligent agent" responsible for the seemingly miraculous emergence of diversity and complexity in the living world. And let's not kid ourselves, Ms. Scott is fond of reminding her audiences with a touch of a smirk, "we're not talking about a disembodied, vague 'intelligence' that 'might' be material. We're talking about God, an intelligent agent that can do things that, according to ID, mortals and natural processes like natural selection cannot."[2] Not for nothing, she points out, does her archenemy Dembski claim that ID is the bridge between science and theology.

Nonsense, Bill Dembski responds, intelligent design is a testable scientific theory—perhaps not all worked out like relativity or quantum mechanics—but eminently scientific nonetheless. The basic approach of employing special mathematical tools to identify evidence of design is common in special sciences like cryptography,[3] SETI research,[4] and archaeology.[5] Why should scientists resist the deployment of such sophisticated design-detection techniques in the field of biology in an honest and straightforward attempt to discern whether an inference of design can be plausibly drawn from the specified complexity exhibited

causes the system to effectively cease functioning."[1]

A living system is thus rather like a Roman arch—each stone is dependent on every other stone for the support of the overall structure. Remove just one stone and the entire edifice will tumble down. How on Earth could such an "irreducibly complex" structure have been assembled in the first place? The obvious answer is that the complicated, mutually dependent system of elements we observe today in the form of finely tuned intracellular machinery masks the history of the evolution of the system—in particular, the presence in the past of scaffolding and crude precursors of one sort or another (vanished biochemical intermediaries and prototypes of modern intracellular processes) that furnished the raw material upon which the process of natural selection could work its magic over the millennia. Behe's analysis simply overlooks the subtlety and power of evolutionary forces operating over vast expanses of time.

by living processes? Isn't it just that the contemporary scientific establishment has a prior intellectual commitment to atheism?

Isn't the preconception on the part of mainstream scientists that God is absent from the world "pure poison"? Dembski continues.[6] On the other hand, doesn't the intelligent design approach hold out the possibility of cleansing our culture of this "poison"? How? By catalyzing a "cultural movement for systematically rethinking every field of inquiry that has been infested by naturalism, reconceptualizing it in terms of design," and, perhaps most important, by launching a "sustained theological investigation that connects the intelligence inferred by Intelligent Design with the God of Scripture and therewith formulates a coherent theology of nature."[7] (I pause to wonder whether even so committed an ideologue as Dembski would seriously contemplate "reconceptualizing" established scientific theories like relativity and quantum mechanics or the standard model of particle physics in order to render those fields of science "connected" with the "God of Scripture.")

Aha, proclaims Eugenie Scott, you have now revealed for all to see that this mumbo-jumbo about design inferences, explanatory filters, and specified complexity is just window dressing. Intelligent design is not a scientific hypothesis at all; it's a Bible-thumping religious movement seeking to accommodate a motley crew of Christian fundamentalists ranging in philosophical orientation from Young Earth Creationism (YEC) to Progressive Creationism (PC) to Theistic Evolutionism (TE). The real clincher, Scott points out, is that IDers dare not suggest a specific alternative to Darwin's dangerous idea for fear of alienating key constituencies within their diverse flock. As Scott puts it, "The reason ID proponents are so vague about an actual picture of what happened is that they strive to include YECs, PCs, and TEs among their theorists and supporters. This is not just a big tent—it is one bulging with people who must be eyeing one another warily."[8]

If intelligent design is ever to attain any level of scholarly respectability, Scott concludes, its proponents are going to have to distinguish their model from the discredited, unscientific ideas of Young Earth Creationism, even if that means losing the support of "biblical literalist

Christians."[9] Instead of attempting to infiltrate anti-Darwinian messages into high school textbooks, IDers who are serious about building a scholarly movement should follow the well-worn path trod by heretical scientific pioneers like the initial proponents of such once outlandish ideas as Big Bang cosmogenesis, plate tectonics, sociobiology, neural Darwinism, punctuated equilibrium evolutionary theory, and the vociferously disdained conjectures of evolutionary theorist Lynn Margulis and others that eukaryotic organelles like mitochondria and chloroplasts are descended from once independent bacteria. The IDers should attend mainstream scientific conferences, submit their papers to recognized peer-reviewed journals like *Nature* and *Science*, Scott counsels, and struggle to promote acceptance of their ideas within the traditional scientific community. Only then will they have a chance of presenting a credible alternative to Darwin's great vision of evolution through random variation and natural selection.

Peace in Our Time?

Is there a chance for peace in our time between these warring factions? And could it be achieved on the terms suggested by Eugenie Scott? Perhaps, but as we shall see, the consequence of "peace" on those terms will likely horrify faithful IDers and ultra-Darwinists alike by both revolutionizing the basic tenets of evolutionary theory and sanitizing ID theory by excising its religious overtones. But does it really matter if the terms of an imposed armistice will be unacceptable to both sides? Is not the mark of a good settlement that it contains something for everyone to hate? And, after all, are we not seeking an *intellectually satisfying* accommodation, not a politically palatable compromise?

With Eugenie Scott's admonition in mind and in the memorable tradition of the negotiators of the Treaty of Versailles (which ended World War I on terms acceptable to absolutely no one), let us begin with the working hypothesis that some version of intelligent design theory might not *supplant* orthodox Darwinism but rather *subsume* it to a more fundamental explanatory paradigm in much the same way that Einstein's theory of relativity subsumes but does not supplant Newton's theory of gravity. Newton's theory continues to live comfortably within the confines of the capacious theoretical edifice erected by Einstein, albeit in a more modest capacity as a limiting case. If intelligent design could legitimately aspire to an analogous objective—to subsume but not supplant Darwinian theory—it would still be undeniably true that, as Dembski puts it, intelligent design could have "radical implications for science."[10]

However, something precious to the IDers must be jettisoned before this effort can begin in earnest: the assumption that intelligent design theory can furnish a bridge linking theology and science. Something abhorrent to the IDers' ideology must be substituted in its place: the clear-headed recognition that nature does not exhibit the slightest trace of the beneficence one might expect of the handcrafted product of a loving, caring Creator. Nature, while in some respects revealing amazing instances of cooperation between species within the biosphere, is in other respects palpably and indisputably red in tooth and claw. Humanity—that great destroyer of other species and of its own kind and perhaps one day of entire worlds—is obviously no exception to the primacy of the law of the jungle. (The ghosts of twentieth-century horrors like Auschwitz stand as irrefutable witnesses to the foregoing proposition.) Contrary delusions from which we may occasionally draw comfort and consolation are precisely that.

If one were to infer the character of a putative supernatural Creator solely from the moral qualities that are exhibited by his or her creation, then one would be forced to the uncomfortable conclusion that we are talking about a demon rather than a deity. At best, the inferred character trait of this hypothetical Creator was aptly described by Richard Dawkins: blind, pitiless indifference to the hapless plight of mankind and other denizens of his or her creation.[11]

This depressing fact is frequently seized upon by opponents of the idea that the universe shows signs of having been intelligently designed as if it were some sort of compelling evidence to the contrary. The physicist and Nobel Prize winner Steven Weinberg, for instance, buttressed his argument against claims that we inhabit a "designer universe" by contending that "there are no signs of benevolence that might have shown the hand of a designer."[12] But why should there be? As even Weinberg admits, "the perception that God cannot be benevolent is very old," citing Euripides, Aeschylus, the Old Testament, and the doctrines of Chistianity and Islam in the same breath.[13] The rhetorical ruse employed by critics like Weinberg seeking to deflect an eminently rational inference of design from the amazingly intricate and life-friendly order of the universe is to shamelessly ridicule the sentimental instincts of the religiously minded, as if such ridicule were a logically persuasive disproof of the design inference. The proper response from design advocates should be to remove this particular debate topic from the ultra-Darwinists' target environment.

The exorcism of religious sentimentality from intelligent design theory is an essential first step as it undercuts utterly the derisive critique offered by orthodox Darwinists that the proponents of intelligent design must be wrong because of their insistence that nature shows signs of the handiwork of a benign creator. More generally, it is the perverse conflation of religion and science that must be thoroughly purged from the ID movement if intellectual progress is to be possible.

An essential second step, as Eugenie Scott points out, is that IDers must strive to construct a solid intellectual model of "what happened" that either contrasts with or, better yet, subsumes Darwinian theory. Lacking such a model, intelligent design doctrine is pathetically open to the accusation leveled by ID critic Frederick C. Crews in the *New York Review of Books* that "intelligent design lacks any naturalistic causal hypotheses and thus enjoys no consilience with any branch of science. Its one unvarying conclusion—'God must have made this thing'—would preempt further investigation and place biological science in the thrall of theology."[14]

But what if a precommitment to supernatural causation were *not* the one unvarying conclusion to emerge from an inquiry into the possibility that the universe was intelligently designed in the distant past? What if a theoretical framework could be constructed that incorporated both a scientifically plausible hypothesis of intelligent design *and* the undeniable fact of terrestrial evolution? And what if such a framework plentifully yielded naturalistic causal hypotheses and enjoyed even greater consilience with branches of science like cosmology and M-theory than does the reigning neo-Darwinian synthesis? Would such an achievement not render Crews's preemptory dismissal of the whole notion of intelligent design a tad premature?

Here's a news flash for Frederick Crews and his ilk: The extraordinarily demanding task of fashioning such a theoretical framework has already begun, not in the shadow of altars and crucifixes but in the conference rooms and lecture halls of path-breaking institutions like the Santa Fe Institute and the Los Alamos National Laboratory. This audacious and frequently derided effort is precisely an attempt to conceive of an alternative model of "what happened." Most succinctly stated, it is an effort to boldly reconceive terrestrial evolution as a subroutine in an inconceivably vast cosmological process of ontogenesis by means of which the universe becomes increasingly pervaded by ever more complex forms of living matter and at the climax of which a living cosmos reproduces itself by propagating one or more "baby" universes.

This attempted reconception of the nature of Nature is emphatically not an attempt to overthrow Darwinism any more than is the Stephen Jay Gould/Niles Eldredge vision of punctuated equilibrium (the notion that speciation events occur at rare intervals, separated by long periods of species stability),[15] Lynn Margulis's notion of species cooperation as an evolutionary strategy,[16] or Stuart Kauffman's idea of self-organization as an engine of evolutionary advancement. On the contrary, it is an effort to take seriously the radical possibility that evolution, in the phrase of superstring theorist Brian Greene, operates on an unexpectedly grand scale[17] and that we can, in the words of Darwinian philosopher Daniel C. Dennett, plausibly "extrapolate a positive Darwinian alternative to the hypothesis that our laws [of physics] are a gift from God"[18] or the inevitable consequence of invariant mathematical rules.

Unlike creationism, this new theoretical effort is utterly indifferent to its religious or sacrilegious implications. The project thus fully meets the challenge posed by Eugenie Wright—to construct an alternative falsifiable scientific model—and strives to achieve seamless consilience with existing bodies of scientific thought, particularly M-theory, cosmology, evolution-through-cooperation notions, and complexity theory.

The framework of the new theory is just now beginning to emerge in peer-reviewed scientific journals like *Complexity* (the journal of the Santa Fe Institute), *Acta Astronautica* (the journal of the International Academy of Astronautics), the *Quarterly Journal of the Royal Astronomical Society*, the *Journal of the British Interplanetary Society*, and other prominent scientific publications. Elements of the theory have been presented and received with great excitement at prestigious scientific conferences like the October 2000 SETI II Session of the International Astronautical Congress and the August 2000 International Artificial Life Conference. Even the distinguished Astronomer Royal of England, Sir Martin Rees, has alluded elliptically to the disconcerting fact that the new theoretical possibility that our own universe was deliberately created by a massively powerful superintellect during a prior cosmic era could resurrect, in radically novel form, the old debate about chance versus design as the underlying theme in the evolution of the cosmos.[19]

The new approach is exceedingly controversial, intensely speculative, and incredibly exhilarating. It is hopefully not egregiously immodest to suggest that this new emerging paradigm has all the earmarks of a major scientific revolution in the making. Why? Because in the words of astrophysicist and science writer Paul Davies quoted in the first chapter, the new paradigm implies that "the laws of nature encode a hidden subtext, a cosmic imperative, which tells them: 'Make life!' And, through life, its by-product: mind, knowledge, understanding. It means that the laws of nature have engineered their own comprehension."[20] How this apparent miracle could have happened, and why, is the subject of the remainder of this book.

Chapter Eight

Cosmic Ontogeny and Earthly Phylogeny

Midway through his entrancing book *Complexification,* mathematician and complexity theorist John Casti inserts a marvelous nugget of insight articulated long ago by the great German philosopher Immanuel Kant. "God," Kant wrote, "has put a secret art into the forces of Nature so as to enable it to fashion itself out of chaos into a perfect world system."[1] Change a few words and Kant's pithy statement could passably describe, albeit in a rather poetic manner, the miraculous process of ontogeny.

Ontogeny, the astonishingly complicated process by which an undifferentiated zygote (a fertilized egg cell) transforms itself into an adult organism, is an intricate waltz of genes and proteins, presided over by an inimitable "maestro": the DNA molecule that uniquely specifies the genome of an embryonic life-form. Interacting networks of feedback loops and molecular switches periodically activate and deactivate particular stretches of the DNA molecule at precise intervals to ensure that embryonic development proceeds smoothly and in the proper sequence.

For thousands of years, as the late evolutionary theorist Stephen Jay Gould pointed out in his imposing treatise *Ontogeny and Phylogeny,* philosophers and scientists have been fascinated by the parallels between ontogeny and phylogeny (the process by which a group of genetically related organisms evolves): "The microcosm: ontogeny. The macrocosm: cosmic history, human history, organic development. This comparison may be the most durable analogy in the history of biology."[2] With this somewhat cryptic comparison Gould attempted to encapsulate the long history of scientific and prescientific speculation about the intriguing parallels between the development of fetal organisms and the birth of new species. Indeed, pre-Socratic philosophers like Anaximander, Anaximenes, and Democritus drew a specific analogy between the embryonic development of an individual organism and the ontogeny of the infant universe. In Gould's words, the "nascent cosmos" was in their cosmological paradigm "surrounded by an envelope resembling the amniotic membrane" and exhibited developmental cycles akin to those of a human embryo.[3]

In Darwin's era, a variation on this ancient theme surfaced. A fascinating notion began to hold sway in the mid–nineteenth century that is associated with the German biologist Ernst Haeckel. His now discredited view can be summed up in a famous slogan: "Ontogeny recapitulates phylogeny." In other words, the developing embryos of "higher" life-forms like humans recapitulate key features of the adult stages of "lower" life-forms from which they presumably evolved. Thus, the presence of gill slits in a developing human embryo was thought to "recapitulate" the gill slits present in adult fishes.

While the specific version of recapitulation advocated by Haeckel has long been discredited, it is nonetheless true, as Gould has trenchantly observed, "[that] some relationship exists cannot be denied."[4] The crucial error committed by Haeckel was his assumption that the ontogeny of a human embryo represents a compressed narrative of the successive appearance of the *adult* forms of distant evolutionary ancestors rather than a developmental pathway that, at early stages, is similar to that of predecessor species. As Gould puts it,

> Haeckel interpreted the gill slits of human embryos as features of ancestral *adult* fishes, pushed back into the early stages of human ontogeny by a universal acceleration of developmental rates in evolving lineages. [An alternative explanation is] that human gill slits do not reflect a change in developmental timing. They are not adult stages of ancestors pushed back into the embryos of descendants; they merely represent a stage common to the early ontogeny of all vertebrates (embryonic fishes also have gill slits, after all).[5]

Under the alternative (and widely accepted) version of the principle, embryos do repeat the embryonic stages of their ancestors, which merely implies that "a conservative principle of heredity" acts to "preserve stubbornly the earlier stages of ontogeny in all members of a [related] group, while evolution proceeds by altering later stages."[6]

The fact that embryonic features of distant evolutionary ancestors are present in the developing fetuses of humans and other late-comers in the Darwinian epic is not the only way in which ontogeny and phylogeny are related. An equally profound connection is furnished by the still controversial hypothesis of neuronal group selection, widely known as neural Darwinism. This hypothesis is the brain child of Nobel laureate Gerald Edelman, who believes that the development of the brain is a selectional process, similar to Darwinian selection, involving populations of neurons engaged in life-and-death competition in the maturing human embryo. The embryonic development of the brain, Edelman suggests, is analogous to the pageant of terrestrial evolution in key respects:

> The anatomical arrangements in the brain and the nervous system are brought about by a series of developmental events. In the embryo, cells divide, migrate, die, stick to each other, send out processes, and form synapses (and retract them). This dynamic series of events depends quite sensitively on place (which other cells are around), time (when one event occurs in relation to another), and correlated activity (whether cells fire together or change together chemically over a period of time).[7]

Like traditional evolution, neuronal group selection yields *diversity* (in the form of specialized cell types) and is driven, at least in part, by *competition* between populations of neurons. While the selectional mechanisms that are responsible for neural development in the embryo obviously differ from those that underlie Darwinian evolution, the important point is that key dynamics of the *phylogenetic* process (competition and selection) are recapitulated in the *ontogenetic* process.

To drive the point home, let us consider the process of neural Darwinism from the viewpoint of an individual nerve cell in the developing embryo. This living entity is literally fighting for its life in a rather hostile environment. If it could conceive of its past, it would recall that its distant ancestors were undifferentiated stem cells that all traced their lineage to a single zygote. If it could somehow gain intelligence about what was happening in the embryo outside of its immediate neighborhood, the tiny nerve cell would realize that a lush and marvelously interconnected ecosystem of many different kinds of "species" of cells was developing, driven in part by competition but in greater measure by the masterful orchestration provided by the DNA molecule shared by the little neuron and all of its fellow cells in the growing embryo. From the nerve cell's perspective, the ontogenetic process would look precisely like Darwinian evolution, with one crucial distinction: the watchmaker would not be blind. On the contrary, the basic repertoire of developmental activities would be subtly and intricately programmed by the embryo's DNA and dynamically modulated by the emergent organization and behavior of the embryo itself.

The nerve cell would realize, if it could think such thoughts, that it was a minuscule participant in a vast self-organizing process of monumental grandeur through which almost endlessly differentiated cell types were being evolved from the simple beginning of a single fertilized egg. The nerve cell would then, if it possessed any semblance of good manners, apologize profusely to Darwin for plagiarizing the magnificent concluding passage of *The Origin of Species*: "There is grandeur in this view of life, with its several powers, having been originally breathed by the Creator into a few forms or into one; and that, whilst this planet has gone cycling on according to the fixed law of gravity, from so simple a beginning endless forms most beautiful and most wonderful have been, and are being evolved."[8] The central point is that what appears superficially to be a process of selection-driven *phylogeny* at one level of analysis may, in fact, constitute a largely preprogrammed process of *ontogeny* at another level of analysis.

To understand how this crucial insight fits into a potential hypothesis explaining the anthropic qualities of our universe, it is important to first ask this key question: How many levels of emergence (the puzzling phenomenon discussed at length in chapter 4) is evolution capable of yielding in principle? And of what feats will evolved life and intelligence be capable after billions of years of future evolution?

Chapter Nine

To the Ends of the Universe

Memes

What is a meme? According to a new definition in the Oxford English Dictionary, a meme (rhymes with cream) is "an element of a culture that may be considered to be passed on by non-genetic means, esp. imitation." The word was coined by Richard Dawkins in his 1976 bestseller The Selfish Gene. *Dawkins was searching for a term that would convey in the context of cultural evolution a meaning closely analogous to that of "gene" in the field of biological evolution.*

The meme concept spread like wildfire. As Dawkins pointed out in a foreword to a 1999 book by evolutionary theorist Susan Blackmore entitled The Meme Machine,[26] *the*

(continued on next page)

It will be recalled that the most troublesome flaw in cosmologist Lee Smolin's daring conjecture of cosmological natural selection was that there was no hypothesized mechanism of heredity. Absent such a mechanism, nothing resembling Darwinian evolution can take place. This defect in Smolin's hypothesis may be characterized as a problem of memory. For purposes of this analysis, I use the term *memory* in the broad sense defined by Gerald Edelman in *Bright Air, Brilliant Fire:*

> I submit that . . . memory . . . takes many forms but has general characteristics that are found in all its variations. I am using the word "memory" here in a more inclusive fashion than usual. Memory is a process that emerged only when life and evolution occurred and gave rise to the systems described by the sciences of recognition. As I am using the term memory, it describes aspects of heredity, immune responses, reflex learning, true learning following perceptual categorization, and the various forms of consciousness. In these instances, structures evolved that permit significant correlations between current ongoing dynamic patterns and those imposed by past patterns. These structures all differ, and memory takes on its properties as a function of the system in which it appears. What all memory systems have in common is evolution and selection. Memory is an essential property of biologically adaptive systems.[1]

Formulation of a hypothesis suggesting a means by which the *memory* (in Edelman's sense) of a particular cosmos could conceivably arise and persist through the process of cosmic replication is the crucial link missing from Smolin's theory. As mathematician John Baez has observed, Smolin's theory is based on two hypotheses:

> A. The formation of a black hole creates "baby universes," the final singularity of the black hole tunneling right on through to the initial "big bang" singularity of the new universe thanks to quantum effects. While this must undoubtedly seem *outré* to anyone unfamiliar with the sort of thing theoretical physicists amuse themselves with these days, in a recent review article by John Preskill on the information loss paradox for black holes, he reluctantly concluded that this was the "most conservative" solution of that famous problem. . . .
>
> B. Certain parameters of the baby universe are close to but different than those of the parent universe. The notion that certain physical facts that appear as "laws" are actually part of the state of the universe has in fact been rather respectable since the application of spontaneous symmetry breaking to the Weinberg-Salam model of electroweak interactions, part of the standard model. . . . So again, while the idea must seem wild to anyone who has not encountered it before, physicists these days are fairly comfortable with the idea that certain "fundamental constants" could have been other than they were. As for the constants of a baby universe being close to, but different than, those of the parent universe, there is as far as I know no suggested mechanism for this. This is perhaps the weakest link in Smolin's argument.[2]

Richard Dawkins made the same point in a public e-mail exchange with Smolin: "Note that any Darwinian theory depends on the prior existence of the strong phenomenon of heredity. There have to be self-replicating entities (in a population of such entities) that spawn daughter entities more like themselves than the general population."[3]

As noted earlier, in the context of Smolin's hypothesis, one can surmise that the physical laws and constants of our universe and its

World Wide Web is literally crawling with meme references. There are numerous online meme discussion groups and at least one peer-reviewed journal focused on "memetics" (the scientific study of memes) as well as a bizarre meme-inspired religion, the "Church of Virus," described by Dawkins as "complete with its own list of Sins and Virtues, and its own patron saint (Saint Charles Darwin, canonized as 'perhaps the most influential memetic engineer of the modern era')"[27]

The memetic perspective—a kind of topsy-turvy point of view where ordinary assumptions about humans' relationship to their culture are perversely inverted—can be disconcerting, to say the least. Evolutionary philosopher Daniel Dennett captures this discomfort nicely: "I don't know about you, but I'm not initially attracted by the idea of my brain as a sort of dung heap in which the larvae of other people's ideas renew themselves, before sending out copies of themselves in an informational Diaspora. It does seem to rob my mind of its importance as both author and critic. Who's in charge, according to this vision—we or our memes?"[28] The memetic viewpoint is disquieting because it regards the transmissible elements of popular culture—advertising jingles, religious beliefs, myths, rumors, etcetera—as the primary actors in the grand drama of cultural evolution, exerting an iron-fisted control over their human hosts precisely analogous to that of Dawkins's "selfish genes" in the pageant of biological evolution.

presumed progeny could conceivably constitute a kind of blueprint (cosmic DNA, if you will) and the universe at large could serve as a sort of mechanism for reading out that blueprint. But what conceivable device or process could play the role of cosmic development controller or cosmic duplicating machine? It may turn out to be the case, as astrophysicist Martin Rees has written, that "[t]he mechanisms that might 'imprint' the basic laws and constants in a new universe are obviously far beyond anything that we can understand."[4] But unless Rees is mistaken and unless Baez's critique can be answered plausibly, Smolin's basic conceptual approach appears to be untenable ab initio.

But could it be that the solution to the puzzle is staring us right in the face? Could life itself and its by-product, biologically evolved intelligence, possibly qualify as the cosmic controller as well as the cosmic duplicator? To begin to answer this daunting question, we must first ask how far the twin processes of evolution (including cultural evolution) and emergence could, at least in principle, propel the phenomenon of complexification in theory. How high could these processes allow mankind and its progeny to achieve progress in intellectual capacity and phenotypic complexity (including the sophistication of what Richard Dawkins would call humanity's extended phenotype in the form of a technological civilization), employing only the ropes and pitons furnished by the principles of complexity theory and evolution? Several recent speculations about the potential magnitude of the ongoing process of cosmological emergence are worth noting as a prelude to formulating a tentative answer to this portentous question.

Future Prospects for the Diffusion of Life Throughout the Universe

The extent to which Earth-based life can, in principle, diffuse throughout the galaxy and beyond has been studied in some detail by astrophysicists. The method of diffusion favored by these theorists is the use of small self-replicating robotic space probes (called "von

Neumann probes") that are able to travel to a nearby stellar system with a minimal payload. The hypothetical probes are programmed to both reproduce themselves for further interstellar journeys and to serve as artificial wombs to permit the growth of frozen human embryos into the living pioneers of the planets where the probes chance to land. (The hypothesis of "directed panspermia," favored by Nobel Prize winner and DNA codiscoverer Francis Crick, suggests that just such a process seeded life on Earth many billions of years ago![5])

According to the theoretical colonization scenario, after an industrial civilization has been established on the first set of alien worlds to be colonized, additional self-replicating probes would be sent to the next set of nearby stellar systems, and so on ad infinitum. Scientists have calculated that a sustained process of galactic colonization, relying on no interstellar propulsion technology more exotic than simple solar sails, could populate the entire Milky Way galaxy with the offspring of earthly life in less than a million years—scarcely the blink of an eye on evolution's long timescale. As astrophysicist Frank Tipler has written,

> [We can] make the reasonable assumption that a von Neumann probe would start making copies of itself within fifty years after reaching the target stellar system. If it sent copies of itself to all stars within ten light-years of itself, colonization of the galaxy could proceed at the rate of ten light-years per sixty years, or 1/6 light-year per year. Since the galaxy is about 100,000 light years in diameter, it would thus take about 600,000 years to colonize the entire Milky Way.[6]

And when will our civilization have the technical capacity to actually begin the colonization process? According to Tipler, by the middle of the twenty-first century![7]

Beyond the Milky Way galaxy loom the vast assemblies of other galaxies, billions in number, separated by dauntingly large intergalactic distances. While these distant galaxies will obviously be more difficult to reach with von Neumann probes than stars in our own Milky Way, Tipler estimates that Earth-originated life could, in principle, engulf the neighboring Andromeda galaxy (2.7 million light-years distant) in approximately 3 million years and the entire nearby cluster of galaxies known as the Virgo Cluster in about 70 million years.[8] Of course, if extraterrestrial civilizations should be present or arise independently in any of these distant locales, the task of diffusing life and intelligence throughout the cosmos becomes considerably simplified.

With this background in mind, let us to proceed to a sober consideration of the theoretical future capabilities of life and intelligence in the cosmos. As we shall see, vast is the loom on which life and intelligence may eventually be capable of reweaving the fabric of reality.

Cosmological Emergence Scenario 1: The Kurzweil Vision

Ray Kurzweil's *The Age of Spiritual Machines* offers a plausible prophecy succinctly conveyed by the subtitle of his book, *When Computers Exceed Human Intelligence.*[9] When exactly will that be? While other artificial-intelligence theorists like Gerald Edelman speculate that we may someday be capable of constructing what Edelman calls a "conscious artifact,"[10] most computer scientists believe that such an extraordinary technological feat lies far in the future.

Not Kurzweil. The computer pioneer forecasts that by 2020, advanced computers will exceed the memory capacity and computational ability of the human brain, with humanlike attributes of synthetic emotion and natural speech not far behind. A mere ten years later, Kurzweil predicts, human brains will be linked seamlessly with their electronic counterparts, allowing information to flow directly between ourselves and our artificial progeny. Not long thereafter, machines will gain decisive advantage over their creators as the inexorable logic of quickening technological innovation (which Kurzweil encapsulates in a general principle he calls the "Law of Time and Chaos") drives their intellectual capacities far beyond ours. As Kurzweil puts it, "Once a computer achieves a human level of intelligence, it will necessarily roar past it."[11]

The debate over whether machines can be endowed with consciousness is scarcely novel. Over the years scientists like Alan Turing and Roger Penrose as well as philosophers like Daniel C. Dennett and John Searle have debated the issue ad nauseam. What is original about Kurzweil's contribution is that he places the anticipated emergence of superior machine intelligence squarely in the context of biological evolution. He states:

> Evolution has been seen as a billion-year drama that led inexorably to its grandest creation: human intelligence. The emergence in the early twenty-first century of a new form of intelligence on Earth that can compete with, and ultimately significantly exceed, human intelligence will be a development of greater import than any of the events that have shaped human history. It will be no less important than the creation of the intelligence that created it, and will have profound implications for all aspects of human endeavor, including the nature of work, human learning, government, warfare, the arts, and our concept of ourselves.[12]

And what are the ultimate prospects for such machines? In Kurzweil's view, they may prove capable of cosmological engineering on the grandest scale:

> [H]ow relevant is intelligence to the rest of the Universe? The common wisdom is, *Not very*. . . . The Universe itself was born in a big bang and will end with a crunch or a whimper; we're not yet sure which. But intelligence has little to do with it. Intelligence is just a bit of froth, an ebullition of little creatures darting in and out of inexorable universal forces. The mindless mechanism of the Universe is winding up or down to a distant future, and there's nothing intelligence can do about it. That's the common wisdom. But I don't agree with it. My conjecture is that intelligence will ultimately prove more powerful than these big impersonal forces. . . . The implication of the Law of Accelerating Returns is that intelligence on Earth and in our

Solar System will vastly expand over time. The same can be said across the galaxy and throughout the Universe. It is likely that our planet is not the only place where intelligence has been seeded and is growing. Ultimately, intelligence will be a force to reckon with, even for these big celestial forces (so watch out!). The laws of physics are not repealed by intelligence, but they effectively evaporate in its presence. So will the Universe end in a big crunch, or in an infinite expansion of dead stars, or in some other manner? In my view, the primary issue is not the mass of the Universe, or the possible existence of antigravity, or of Einstein's so-called cosmological constant. Rather, the fate of the Universe is a decision yet to be made, one which we will intelligently consider when the time is right.[13]

Cosmological Emergence Scenario 2: The Wheeler Vision

John Wheeler's vision of the distant future, which has been dubbed the participatory anthropic principle, was summarized in the second chapter of this book. In essence, Wheeler believes that all of reality has a computational origin—"it from bit"—and indeed that reality is actually summoned into existence by countless acts of observer-participancy by hordes of conscious observers throughout the universe, most of whom inhabit the far future!

Wheeler's vision seeks to literally extend the counterintuitive implications of observer-participancy evidenced in quantum physics experiments to the physical and temporal ends of the cosmos. In his vision, the aggregate impact on the physical state of the universe of *future* acts of observer-participancy is decisively greater than the impact of current acts of observer-participancy for two reasons. First, future observers are likely to outnumber the current population of observer-participants. With billions of stellar systems remaining for humanity to explore and possibly colonize, this assumption is difficult to dispute as an abstract proposition. Second, the impact of future acts of observer-participancy is likely to be qualitatively different (and more profound) because of the emergence of new modalities by which life and intelligence will participate in the future in measuring (and thus shaping) the cosmos. Wheeler explains:

> More is different. . . . We do not have to turn to objects so material as electrons, atoms, and molecules to see big numbers generating new features. The evolution from small to large has already in a few decades forced on the computer a structure reminiscent of biology by reason of its segregation of different activities into distinct organs. Distinct organs, too, the giant telecommunications system of today finds itself inescapably evolving. Will we someday understand time and space and all the other features that distinguish physics—and existence itself—as the similarly self-generated order of a self-synthesized information system?[14]

Wheeler has committed himself unambiguously to the proposition that the laws and constants of inanimate nature are not static but highly plastic and that they are given form and content by the retroactive quantum mechanical impact of many intelligent observers processing data about those processes and their outcome. This is precisely what Wheeler means when he asserts that we inhabit a "participatory" universe. In such a scenario, the anthropic quality of the cosmos follows as a matter of course from the initial proposition about the key role of observer-participancy in generating reality.

Cosmological Emergence Scenario 3: The Barrow/Tipler Vision

The Barrow/Tipler Omega Point theory of the cosmos, dubbed the final anthropic principle, has already been summarized in detail. The important point to emphasize is that the essence of this theory is its prediction that life and life-mediated processes will, in the distant future, be able to gain total and indisputable control over inanimate matter and the great impersonal forces that today power the process of cosmological evolution. Barrow and Tipler conceive of one possible exotic "reaso[n] to think that life is essential to the Cosmos"[15] (which would be to keep the universe from destroying itself by preventing black hole evaporation!) but do not address the possibility that the life processes might eventually be capable of the ultimate feat of cosmological engineering: cosmic replication.

Cosmological Emergence Scenario 4: The Dyson Vision

In *Infinite in All Directions* Freeman Dyson offers a vision of the distant future similar in many respects to the Barrow/Tipler scenario. Like the preceding speculators, Dyson places life and intelligence at the center of any serious inquiry into the ultimate fate of the cosmos:

> It is impossible to calculate in detail the long-range future of the universe without including the effects of life and intelligence. It is impossible to calculate the capabilities of life and intelligence without touching, at least peripherally, philosophical questions. If we are to examine how intelligent life may be able to guide the physical development of the universe for its own purposes, we cannot altogether avoid considering what the values and purposes of intelligent life may be.[16]

Dyson contrasts his vision with the nihilistic observation of Steven Weinberg quoted previously ("the more the universe seems comprehensible, the more it also seems pointless"), foreseeing a universe ever more suffused with life, intelligence, and purpose:

> The universe that I have explored in a very preliminary way . . . is very different from the universe which Weinberg envisaged when he

> called it pointless. I have found a universe growing without limit in richness and complexity, a universe of life surviving forever and making itself known to its neighbors across unimaginable gulfs of space and time. Whether the details of my calculations turn out to be correct or not, there are good scientific reasons for taking seriously the possibility that life and intelligence can succeed in molding this universe of ours to their own purposes.[17]

Dyson even divines a candidate "law of nature" from the tendency of conscious thought to exert ever greater control over inanimate matter:

> To me the most astounding fact in the universe . . . is the power of mind which drives my fingers as I write these words. Somehow, by natural processes still totally mysterious, a million butterfly brains working together in a human skull have the power to dream, to calculate, to see and to hear, to speak and to listen, to translate thoughts and feelings into marks on paper which other brains can interpret. Mind, through the long course of biological evolution, has established itself as a moving force in our little corner of the universe. Here on this small planet, mind has infiltrated matter and has taken control. It appears to me that the tendency of mind to infiltrate and control matter is a law of nature.[18]

The operation of this "law of nature," Dyson believes, implies that life and intelligence will play a dominant role in shaping the physical eschatology of the cosmos:

> Individual minds die and individual planets may be destroyed. But, as Thomas Wright said, "The catastrophe of a world, such as ours, or even the total dissolution of a system of worlds, may possibly be no more to the great Author of Nature, than the most common accident of life with us." The infiltration of mind into the universe will not be permanently halted by any catastrophe or by any barrier that I can imagine. If our species does not choose to lead the way, others will do

> so, or may have already done so. If our species is extinguished, others will be wiser or luckier. Mind is patient. Mind has waited for 3 billion years on this planet before composing its first string quartet. It may have to wait for another 3 billion years before it spreads all over the galaxy. I do not expect that it will have to wait so long. But if necessary, it will wait. The universe is like a fertile soil spread out all around us, ready for the seeds of mind to sprout and grow. Ultimately, late or soon, mind will come into its heritage.[19]

What use will life and intelligence make of this "fertile soil"? Dyson is deeply skeptical about the capacity of our inherently bounded human intellects to probe this ultimate mystery:

> What will mind choose to do when it informs and controls the universe? This is a question which we cannot hope to answer. When mind has expanded its physical reach and biological organization by many powers of ten beyond the human scale, we can no more expect to understand its thoughts and dreams than a Monarch butterfly can understand ours. . . . In contemplating the future of mind in the universe, we have exhausted the resources of our puny human science. This is the point at which science ends and theology begins.[20]

But must this necessarily be the point at which science ends? Or could it merely mark the frontier of current theory—like the edge of an ancient map showing the outer boundaries of explored territories and displaying images of dragons and monsters beyond the edge of the known world?

Cosmological Emergence Scenario 5: The Dawkins Vision

The work of evolutionary theorist Richard Dawkins is not ordinarily associated with cosmological models or philosophizing about the ultimate fate of the universe. Nonetheless, his notion of ascending hierarchies of replicators pursuing emergent categories of replication objectives furnishes a valuable conceptual tool with which to synthesize the insights of the four theoreticians discussed above. Dawkins's vision of the hierarchical layering of emergent replicator categories is stated most clearly in the concluding chapter of his book *River Out of Eden*:

> There is another type of explosion [besides a supernova explosion] a star can sustain. Instead of "going supernova" it "goes information." The explosion begins more slowly than a supernova and takes incomparably longer to build up. We can call it an information bomb or, for reasons that will become apparent, a replication bomb. For the first few billion years of its build-up, you could detect a replication bomb only if you were in the immediate vicinity. Eventually, subtle manifestations of the explosion begin to leak away into more distant regions of space and it becomes, at least potentially, detectable from a long way away. We do not know how this kind of explosion ends. Presumably it eventually fades away like a supernova, but we do not know how far it typically builds up first. Perhaps to a violent and self-destructive catastrophe. Perhaps to a more gentle and repeated emission of objects, moving, in a guided rather than a simple ballistic trajectory, away from the star into distant reaches of space, where it may infect other star systems with the same tendency to explode.[21]

Human life, in Dawkins's view, plays an important catalytic role in the "detonation" of the replication bomb. He explains, "We humans are an extremely important manifestation of the replication bomb, because it is through us—through our brains, our symbolic culture and our technology—that the explosion may proceed to the next stage and reverberate through deep space."[22]

The commencement of the process, however, antedates the arrival of the human race by billions of years:

> We have no direct evidence of the replication event that initiated the proceedings on this planet. We can only infer that it must have happened because of the gathering explosion of which we are a part. We do not know exactly what the original critical event, the initiation of self-replication, looked like, but we can infer what kind of an event it must have been. It began as a chemical event. . . . What, then, was this momentous critical event that began the life explosion? I have said that it was the arising of self-duplicating entities, but equivalently we could call it the origination of heredity—a process of "like begets like."[23]

Dawkins proposes a replicator classification system that comprises a detailed hierarchy of "replicator thresholds," beginning with Threshold 1 (the "Replicator Threshold" itself)[24] continuing through Threshold 5 (the "High-Speed Information Processing Threshold"), Threshold 6 (the "Consciousness Threshold"), Threshold 7 (the "Language Threshold"), and Threshold 8 (the "Cooperative Technology Threshold").

At some point between Threshold 5 and Threshold 8, Dawkins theorizes, an entirely new class of replicators arose: selfish, self-replicating memes. He states, "[I]t is possible that human culture has fostered a genuinely new replication bomb, with a new kind of self-replicating entity—the meme, as I have called it in *The Selfish Gene*—proliferating and Darwinizing in a river of culture. There may be a meme bomb now taking off, in parallel to the gene bomb that earlier set up the brain/culture conditions that made the take-off possible.[25] (See sidebar, page 103.)

The final replication threshold foreseen by Dawkins is Threshold 10 (the "Space Travel Threshold"), which he describes as follows:

> After radio waves, the only further step we have imagined in the outward progress of our own explosion is physical space travel itself: Threshold 10, the Space Travel Threshold. Science-fiction writers have dreamed of the interstellar proliferation of daughter colonies of humans, or their robotic creations. These daughter colonies could be

> seen as seedlings, or infections, of new pockets of self-replicating information—pockets that may subsequently themselves expand explosively outward again, in satellite replication bombs, broadcasting both genes and memes. If this vision is ever realized, it is perhaps not too irreverent to imagine some future Christopher Marlowe reverting to the imagery of the digital river: "See, see, where life's flood streams in the firmament!"[29]

The momentous question posed by Lee Smolin's daring conjecture of cosmological natural selection can be restated in terms of Dawkins's classification scheme: Is Threshold 10 truly the final replication threshold? Or might there be a Threshold 11, which we may provisionally call the "Cosmic Replication Threshold"? Might Threshold 11 harbor a radically new type of replicator—differing from the preceding classes as profoundly as the meme differs from the gene but incorporating the complex interactions of those precedent entities as subroutines—that we might provisionally label (in deference to Dawkins's memorable nomenclature) as the "Selfish Biocosm" replicator class?

Synthesis of the Five Cosmological Emergence Scenarios

We are now prepared to formulate a crude synthesis of key elements of the five cosmological emergence scenarios discussed above. This synthesis yields two dramatic hypotheses.

- First, according to the first four scenarios, life is capable of attaining the capacity to engage in cosmological engineering in the very distant future. While the mechanism postulated by Wheeler (retroactive quantum mechanical effects of observer-participancy) differs from the vaguely stated technological assumptions of the other theorists, all agree on the potential magnitude of the future effect of life on the global state of the cosmos.

- Second, Dawkins's open-ended replicator hierarchy suggests that the natural processes of self-organization, emergence, and natural selection, governed by laws whose existence is hypothesized (but not yet definitively formulated) by complexity theorists as well as by theoretical approaches derived from Darwinian theory, are fully capable of yielding such a capability without any requirement of supernatural intervention or supervision.

As we shall see, these two key hypotheses provide the foundation for a startling new vision of the central role potentially played by life and intelligence in the birth, evolution, and replication of the universe. It is a vision at once disconcerting and thrilling, intimidating and awe inspiring.

Part Four

Point Omega: Dreams of a Transhuman Rapture

Part 4 is the heart of this book: an articulation of the Selfish Biocosm hypothesis. The basic idea is that the anthropic, or life-friendly, qualities that our universe exhibits are logical and predictable consequences of a cosmic reproduction cycle in which a cosmologically extended biosphere, developed and evolved over billions of years to unimaginable levels of sophistication, serves as the device by which our cosmos duplicates itself and propagates one or more "baby universes." The hypothesis portrays the cosmos as "selfish" in the same metaphorical sense that Richard Dawkins proposed that genes are "selfish." Under the Selfish Biocosm theory, the cosmos is "selfishly" focused upon the overarching objective of assuring its own replication. As economists would put it, self-reproduction is the hypothesized "utility function" of the universe.

BIOCOSM

Chapter Ten

The Selfish Biocosm: An Emergent Replicator Class

Michael Denton: Defender of the Anthropocentric Faith

Dr. Michael Denton is a researcher in the biochemistry department in the University of Otago in New Zealand. His main research interest is the molecular genetics of human retinal disease. The author of over seventy scientific papers, Denton achieved notoriety with his books about evolution (Evolution: A Theory in Crisis) *and the anthropic principle (Nature's Destiny: How the Laws of Biology Reveal Purpose in the Universe). The latter is a lengthy brief in favor of the proposition that human beings (or, at the very least,*

(continued on next page)

We began the last chapter with the goal of determining whether the sciences of complexity could come to the rescue of Lee Smolin's daring conjecture of cosmological natural selection by supplying two indispensable components conspicuously absent from his theory: a controller and a duplicating machine. These two components, it will be recalled, have been shown by John von Neumann to be logically essential elements of *any* self-reproducing entity. The omission of the second element is particularly glaring because any hypothesized process that involves evolution through natural selection obviously requires a robust mechanism of heredity or, in Gerald Edelman's phrase, a capacity for high-fidelity memory. Synthesis of the five cosmological emergence scenarios summarized in the previous chapter together with an admittedly speculative application of key concepts underlying the sciences of complexity yields a preliminary answer to the question presented by the first sentence above:

1. Tentatively identified principles guiding the evolution and operation of complex adaptive systems could conceivably function as a controller, governing the process of cosmological self-organization and emergence at all relevant scales, leading up to a cosmological replication event.

2. Life itself, when it reaches a requisite threshold of pervasiveness and evolved sophistication at or near the Barrow/Tipler Omega Point, could conceivably serve as the requisite cosmological duplicating machine.

This startling conclusion is the essence of the Selfish Biocosm hypothesis. Let us pause for a moment to appreciate and savor the revolutionary nature of the claims stated above.

Freeman Dyson, contemplating the incredible set of coincidences that had to exist in order for intelligent life to arise in the universe, offered this trenchant observation: "The more I examine the universe and study the details of its architecture, the more evidence I find that the universe in some sense must have known that we were coming. There are some striking examples in the laws of nuclear physics of numerical accidents that seem to conspire to make the universe habitable."[1]

Dyson was referring to the oddly fine-tuned physical constants and laws of nature that various versions of the anthropic cosmological principle seek to explain. His speculation was that the universe seemed somehow to have been set up from the very beginning to eventually give birth to life and intelligence. This teleological explanation seems improbable in the extreme (absent the deus ex machina of a designing divinity) *unless* the phenomenon of cosmic origin *and* the process of biological evolution are causally linked in a very particular way.

The traditional interpretation of the set of coincidences comprising the anthropic principle is to assert that they constitute a selectional constraint on universes capable of yielding life and intelligence. This assertion is certainly true as far as it goes, but it hardly qualifies as a satisfying causal explanation.

In contrast, the novel interpretation suggested by the Selfish Biocosm hypothesis is that the life-friendly quality of the physical laws that dominate our cosmos is a causal and fully naturalistic consequence of the fact that highly evolved life and intelligence constitute the duplicating machine that is responsible for the replication and re-creation of universes like ours. In other words, it is not that a vastly improbable set of cosmic coincidences somehow managed to come together fortuitously to randomly endow our universe with its peculiar life-giving qualities; it is rather that highly evolved life (and its by-product intelligence) is the hypothesized causal agent that gave birth to our universe and, in the

creatures closely resembling humans) were the intended culmination of a teleogically focused process of biological evolution. As Denton puts it,

The cosmic "telos" I have in mind is advanced carbon-based humanlike or humanoid life. It is not specifically our own unique species Homo sapiens. At present, there is insufficient evidence to argue that the laws of nature are uniquely fit for every detail of human biology exactly as found in our own species today. However, I believe that the current evidence points strongly in this direction and that future scientific advances will confirm the absolute centrality of mankind in the cosmic scheme.[11]

To his credit, Denton forthrightly acknowledges that this unapologetically anthropocentric point of view—that the appearance of our own species is the fundamental end and purpose of the entire process of cosmic evolution—would be utterly shattered if there ever emerged, anywhere in the cosmos, a form of artificial intelligence superior to that of humankind. He states: "The creation of a machine with an intellectual capacity superior to that of man would . . . effectively demolish the argument that the universe is contrived with mankind and human intelligence as its primary end."[12]

process, replicated the full suite of life-giving physical qualities of a predecessor cosmos, thereby transmitting to our universe the same life-mediated reproductive capability possessed by the "mother" universe. Under this scenario, the life-friendly physical attributes of the cosmos and the puzzling phenomena of the origin and evolution of life and intelligence are tightly linked in a hypothesized causal relationship.

Let us now examine the conjecture in more detail and begin to ask whether it constitutes a testable scientific hypothesis or is merely a metaphysical or quasi-religious speculation. Following are seven key questions to help us resolve that issue.

Question #1: A Natural or Supernatural Scenario?

Does the Selfish Biocosm scenario depict a natural or a supernatural scenario? If the latter, the scenario has no place in the ample abode of science. Clearly, the scenario depends solely on hypothesized natural processes. What is startling is not the basic nature of those processes but rather the potential magnitude of their predicted eventual impact over immense stretches of cosmic time—namely, the pervasive colonization of inanimate matter by life and the attainment by life and intelligence of the capacity to engage in the ultimate feat of cosmological engineering (creation of a baby universe)—on the global state of the cosmos. But that does not mean that the hypothesized processes are any less natural than the inscrutable processes that gave birth to the Big Bang or to living matter or, more pertinently, to the presumptively naturalistic processes that underlie the ongoing phenomenon of biological and cultural evolution and complexification.

It may be, as Darwin once concluded with respect to deep questions about the cosmos and the mystery of life's origin, "that the whole subject is too profound for the human intellect. A dog might as well speculate on the mind of Newton."[2] But that is merely a reflection of the limitations of the human mind that evolution has thus far bequeathed

How to Build a Baby Universe

The exotic topic of how to build a baby universe has received a surprising amount of attention in the scholarly literature. This esoteric discussion is motivated by the realization that a popular variant of standard cosmic inflation theory—eternal chaotic inflation, a hypothesis pioneered by Russian-born cosmologist Andrei Linde—holds that the Big Bang was not a

(continued on next page)

to us. It is not a principled objection to the utterly *natural* character of the proposed scenario.

Question #2: Does the Scenario Offer an Alternative Narrative of "What Happened"?

Eugenie Scott's criterion for a serious alternative to the blind watchmaker theory of unguided natural selection is that the proposed hypothesis must offer a compelling narrative of what happened. Here the narrative is compelling, albeit unconventional. It depicts an almost unthinkably vast and lengthy process through which a particular universe is born from the loins (metaphorically speaking) of a predecessor cosmos; undergoes an initial phase of superluminal inflation and lifeless self-organization; eventually yields living creatures; spawns the phenomenon of biological evolution and the subsequent emergence of intelligence and ensuing cultural evolution; witnesses the gradual expansion of the living domain until life's dominion becomes coterminous with the outer ramparts of the cosmos itself; and then, through the instrumentality of a supremely evolved life-form provisionally called the Selfish Biocosm, reproduces itself and creates a new baby universe.

It is no proper objection to this proffered scenario of what happened that the required scale of future evolutionary time is, from our human perspective, inconceivably vast. It must not be forgotten that Darwin's own monumental achievement provides ample reason to believe that limitations on our evolved intelligence are not insurmountable obstacles to human comprehension of even the most cosmically vast and counterintuitive physical phenomena. Answering the question of why the most eminent geologists and naturalists had, until shortly before publication of *The Origin of Species*, disbelieved in the mutability of species, Darwin responded that this false conclusion was "almost unavoidable as long as the history of the world was thought to be of short duration."[3] It was geologist Charles Lyell's speculations on the immense

singular, one-off event but rather was exemplary of an endless firecracker storm of Big Bangs that have exploded and will continue to explode in unobservable portions of our universe (and others) for all of eternity.

The question naturally arises: If nature can create new universes with such apparent ease and in such profusion, could scientists replicate that feat in the laboratory, at least in principle? The answer appears to be a tentative "maybe."

The most promising approach, explored in scientific papers by Linde and other prominent scientists, seems to be a technique rooted in quantum theory. Scientists would use a carefully controlled application of a well-known quantum mechanical phenomenon—quantum tunneling—to propel a tiny bubble of false vacuum out of the bounds of our universe and coax it into a state of exponential expansion, which would result in a new Big Bang. As cosmologist Alan Guth points out, further exploration of this exotic possibility depends on a complete theory of quantum gravity, which does not presently exist.[14]

age of Earth that provided the essential conceptual framework for Darwin's own theory. Lyell's vastly extended stretch of geological time provided an ample temporal arena in which the forces of natural selection could sculpt and reshape the species of Earth and achieve nearly limitless variation.

The central point is that collateral advances in sciences seemingly far removed from biology can dissipate the intellectual limitations imposed by common sense and naive human intuition. In our era, the key advances provided by physics (quantum theory, relativity, and M-theory) as well as cosmology (the Big Bang hypothesis and the newer contributions discussed subsequently) can assist biologists in precisely the manner that Lyell's new geology provided critical assistance to Darwin. All that is needed is for biologists to take seriously the counterintuitive implications of these new theoretical advances.

In particular, attention must be paid to the key insight articulated by John Wheeler and many other physicists: most of the time available for life and intelligence to achieve their ultimate capabilities lie in the cosmic future, not in the cosmic past. As cosmologist Frank Tipler bluntly stated, "Almost all of space and time lies in the future. By focusing attention only on the past and present, science has ignored almost all of reality. Since the domain of scientific study is the whole of reality, it is about time science decided to study the future evolution of the universe."[4]

Louis Crane's Meduso-Anthropic Principle

If you scan the acknowledgment page of astrophysicist Lee Smolin's 2001 book, Three Roads to Quantum Gravity, *you will notice, prominently displayed, the unfamiliar name of Louis Crane.*[17] *Kansas State University mathematician Crane toils quietly at the task of formulating a plausible theory of quantum gravity in the unlikely venue of Manhattan, Kansas.*

Crane is not only a gifted quantum-gravity theorist but also an imaginative cosmologist of the first order who has come up with a unique twist on "baby universe" creation

(continued on next page)

Question #3: Is the Scenario Consilient with Other Fields of Science?

The Selfish Biocosm hypothesis enjoys profound—indeed, increasingly compelling—consilience with other branches of science, including quantum theory, astrophysics, M-theory, and theories about the nature of evolutionary progress through deep symbiosis proffered by Lynn Margulis and others. In particular, the hypothesis is deeply consilient with Darwin's own theory and with the new sciences of complexity. The hypothesis relies upon the assumption that the natural phenomena of evolution (including cultural evolution) and emergence are capable, without divine intervention or direction, of driving life and intelligence upward to an almost unimaginably high level of sophistication, complexity, and capability in the distant future.

Question #4: Does the Scenario Offer a More Encompassing Explanation Than Darwin's?

It is important to be clear on one key point. The Selfish Biocosm hypothesis does not purport to supplant or overthrow Darwin's great theory but rather to *subsume* it to a more encompassing explanation of "what happened." Under the proffered scenario, terrestrial evolution is reconceptualized as a subroutine in an inconceivably vast ontogenetic process through which the universe gives birth to life and intelligence and, through the mechanism of highly evolved versions of those intermediaries, is able to reproduce itself. The scenario offers a more encompassing explanation than Darwin's theory because it attempts to explain the essential qualities of inanimate as well as animate matter and to link those explanations. Indeed, it proposes to conceptually unite these two seemingly incommensurable realms by asserting that the oddly life-friendly laws and physical constants of inanimate nature are life-friendly *precisely because* they serve a function analogous to DNA in an earthly creature: they collectively provide a *recipe* for cosmic ontogenesis (a process that includes the birth of life and biological evolution) and a *blueprint* for the construction of cosmic offspring. In this respect the scenario is consonant with Harold Morowitz's notion that the origin of life was not an accident but rather was causally rooted in the basic laws of chemistry[5] and the complementary idea expressed by Christian de Duve that "life arose naturally, by the sole enactment of physical and chemical laws."[6]

theory. According to Crane's "Meduso-Anthropic Principle," the universe is delicately tuned for the production of life and intelligence, not black holes (as Smolin proposes). As under Smolin's scenario, Crane suggests that black holes seed new baby universes and the baby universes bear a family resemblance to their parents (i.e., the rules of physics in the offspring are closely similar to those that prevail in the "mother" universe). However, Crane departs from Smolin by speculating that artificially created black holes will proliferate in the future.

Why would advanced civilizations bother to manufacture black holes? Not out of altruistic motives, according to Crane, but as a simple expedient for generating energy in an aging universe facing the prospect of heat death![18]

Question #5: Does the Scenario Promote or Demote the Cosmic Role of Humanity?

According to the late Stephen Jay Gould, "Sigmund Freud argued that all great scientific revolutions feature two components: an intellectual reformulation of physical reality and a visceral demotion of *Homo sapiens* from arrogant domination atop a presumed pinnacle to a particular and contingent result, however interesting and unusual, of natural processes."[7] Freud pointed to two such revolutions as exemplary: the "Copernican banishment of Earth from center to periphery and the Darwinian 'relegation' . . . of our species from God's incarnated image to 'descent from an animal world.'"[8]

Does the Selfish Biocosm promote or demote the role of humanity in the cosmos? Oddly, it does both simultaneously. The hypothesis unquestionably elevates the role of life and intelligence in general in the proposed process of cosmic ontogeny and replication. However, this elevation scarcely implies that the universe exists to serve the interests of mankind; if anything, it suggests precisely the contrary.

From the Selfish Biocosm perspective, earthly life and human intelligence are not the grand climax of creation but rather minuscule operants in a surpassingly complex process that our particular universe employs in order to get itself grown to maturity and then reproduced. The disquieting jolt induced by this change of perspective is reminiscent of that furnished by the "meme's eye view" of human culture, characterized as follows by Daniel C. Dennett:

> This [memetic perspective] is a new way of thinking about ideas. It is also, I hope to show, a good way, but at the outset the perspective it provides is distinctly unsettling, even appalling. We can sum it up with a slogan: A scholar is just a library's way of making another library.[9]

To view the multibillion-year pageant of life's proliferation on Earth and the evolution of human intelligence as minor subroutines subordinated to inconceivably vast ontogenetic processes through which our particular universe prepares itself for replication is scarcely to place mankind at the center of creation. It is rather to adopt a profoundly super-Copernican perspective (to use John Wheeler's phrase). Far from offering an anthropocentric vision of the cosmos, the Selfish Biocosm perspective relegates humanity and its probable mechanical progeny to the functional equivalents of mitochondria—formerly independent biological entities whose talents were harnessed in the distant past to serve the greater good of eukaryotic ascendance. (Eukaryotic cells are large complex cells that include separate nuclei containing DNA and tiny organelles, some of which—like mitochondria—are thought to have formerly been free-living bacteria whose talents—like energy generation and photosynthesis—were harnessed by the eukaryotic cells eons ago and which then took up permanent residence inside the larger cells as endosymbionts.) In this respect, the hypothesis should be sharply distinguished from forthrightly anthropocentric cosmological scenarios like scientist and intelligent design advocate Michael Denton's religiously in- spired *Nature's Destiny*, which offers the idiosyncratic view that the laws of nature are fine-tuned to reach an evolutionary endpoint in mankind.[10] (See sidebar, page 121.)

Question #6: Does Any Serious Scientist Believe That a Baby Universe Can Be Fabricated?

Alan H. Guth is a physics superstar. He is V. F. Weisskopf Professor of Physics at the Massachusetts Institute of Technology and the chief proponent of the cosmological theory known as "inflation"—the idea that the infant universe underwent a stage of ultra-rapid expansion shortly after the Big Bang.

As he recounts in his book *The Inflationary Universe*, Guth has actually undertaken a serious study of whether a new baby universe could, at least in theory, be produced "in the lab" by a sufficiently advanced life-form:

> Since the inflationary theory implies that the entire observed universe can evolve from a tiny speck, it is hard to stop oneself from asking whether a universe can in principle be created in a laboratory. Given what we know of the laws of physics, would it be possible for an extraordinarily advanced civilization to play the role of Doondari or Marduk, creating new universes at will? In collaboration with several graduate students and colleagues at MIT, I began to study this question in the mid-1980s.[13]

Guth's surprising conclusion is that it may, in fact, be possible in principle for a superadvanced civilization to create an artificial baby universe in the lab, although the technology required is probably billions of years in the future. (See sidebar, page 123.)

Perhaps most important with respect to the credibility of Guth's prognostication is that while the technological means to create a baby universe is not currently within our grasp, the scientific principles underlying such a future technical capability are simple to state within the confines of present-day science. As Kansas State University mathematician Louis Crane has explained in an unpublished paper,

> Can an advanced industrial civilization create small hot black holes [which would serve as gateways to artificially created baby universes]? It is of course not possible to address this at the level of a practical engineering question at the present time. As far as basic science goes, the answer is a clear yes. In order to create a black hole, it is necessary to concentrate a sufficiently large amount of mass/energy in a sufficiently small region of space. Bosonic matter, such as electromagnetic radiation, is the natural means to attempt such a thing, since it is not subject to the [Pauli] exclusion principle. To accomplish the goal, it is necessary to fire a sufficiently large sphere of lasers inwards at a

> sufficiently simultaneous time. A civilization that had at its disposal a sufficiently large numbers of automated factories in outer space should be able to assemble such a sphere. Once the sphere was built, it could go on producing black holes as quickly as it could be re-excited. . . . If it were possible to build nuclear lasers, the sphere would be the mass of a small asteroid.[15]

In short, the Selfish Biocosm hypothesis is completely consistent with the prevailing scientific consensus regarding the question of whether there are foreseeable, though not yet feasible, engineering techniques based on currently available scientific concepts that could possibly be employed in the distant future to artificially create a baby universe and imprint it with the laws of physics that prevail in our own universe. As astronomer Martin Rees put it in his 1994 book, *Before the Beginning: Our Universe and Others,*

> Some cosmologists speculate that new "embryo" universes can form within existing ones. Implosion to a colossal density (around, for instance, a small black hole) could trigger the expansion of a new spatial domain inaccessible to us. Universes could even be "manufactured"—the experimental challenge is far beyond present human resources, but may become feasible, especially if we recall that our universe has most of its course still to run. No information could be exchanged with a daughter universe, but it would bear the imprint of its parentage. Our own universe might be the (planned or unplanned) outcome of such an event in some preceding cosmos. The traditional theological "argument from design" then reasserts itself in novel guise.[16]

A few daring scientists, including Louis Crane, have even speculated about the motivations that might induce sufficiently advanced civilizations to manufacture black holes and baby universes "in the lab." (See sidebar, page 125.)

"What if Darwin's principle of natural selection were merely a tiny fractal embodiment of a universal life-giving principle that drives the evolution of stars, galaxies, and the cosmos itself? What if the universe were literally in the process of coming to life?"

—James N. Gardner

Stellar "Fireworks Finale" Came First in the Young Universe

An artist's impression of how the very early universe (less than 1 billion years old) might have looked when it went through a voracious onset of star formation, converting primordial hydrogen into myriad stars at an unprecedented rate. Back then, the sky would have looked markedly different from the sea of quiescent galaxies around us today.

Hubble Images Swarm of Ancient Stars

One of the densest of the 147 known globular star clusters in the Milky Way galaxy contains hundreds of thousands of stars, all held together by their mutual gravitational attraction. These are all about 15 billion years old, and every star visible is more highly evolved than our own Sun. The bright red giants are stars similar to the Sun in mass that are nearing the ends of their lives.

Hubble Reveals the Heart of the Whirlpool Galaxy

The Whirlpool Galaxy is having a close encounter with a nearby companion galaxy, just off the upper edge of this image, whose gravitational pull is triggering star formation in the main galaxy, as seen in brilliant detail by numerous, luminous clusters of young and energetic stars. The bright clusters are highlighted in red by their associated emission from glowing hydrogen gas.

The Crab Nebula

This is a remnant of a star 10 times the mass of our own Sun that exploded as a supernova on July 4, 1054. The various colors arise from different chemical elements in the expanding gas, including hydrogen (orange), nitrogen (red), sulfur (pink), and oxygen (green). Astronomers believe that the chemical elements in the Earth and even in our own bodies were made in other exploding stars billions of years ago.

Colorful Fireworks Finale Caps a Star's Life

The colorful streamers were created by the titanic supernova explosion of a massive star. The debris is arranged into thousands of small, cooling knots of gas. This material eventually will be recycled into building new generations of stars and planets. Our own Sun and planets are constructed from the debris of supernovae that exploded billions of years ago.

Multiple Generations of Stars in the Tarantula Nebula

Many of the stars in this region are so old that they have exploded as supernovae, blasting material out into the surrounding region at speeds of almost 200 miles per second. Near the center of the image are small, dense gas globules and dust columns where new stars are being formed today, as part of the overall ongoing star formation throughout the Tarantula Nebula.

Question #7:
Is the Selfish Biocosm Hypothesis Testable?

The acid test for determining whether an idea qualifies as a scientific hypothesis as opposed to a mere metaphysical speculation is whether or not it yields falsifiable implications. Can the Selfish Biocosm concept qualify as a genuine scientific hypothesis under this standard? That depends on whether it generates falsifiable implications. As we shall see in the next chapter, it does so in abundance.

Chapter Eleven

Testing the Selfish Biocosm Hyopthesis: Rendevous in Rio

Artificial Life

The new science of artificial life is a child of the computer era. It rests on the hypothesis that the biochemical processes that energize living matter are actually elaborate forms of computation that are susceptible to simulation in conventional computers. Nurtured in centers of complexity science like the Los Alamos National Laboratory, the Sante Fe Institute, and MIT, artificial life investigation has become a thriving scientific specialty, boasting well-attended international confer-ences and impressive peer-reviewed journals exclusively dedicated to the field. Even NASA has become, somewhat belatedly, an "a-life" cheerleader, promising to include the new

(continued on next page)

The transcendently beautiful city of Rio de Janeiro was the site of the Fifty-first International Astronautical Congress, convened the austral springtime of October 2000. The congress took place at the RioCentro Convention Center, site of the famous 1992 Earth Summit, which fo-cused the attention of the world on the effect of human technological civilization on the natural environment. I was in Rio at the invitation of Jill Tarter, head of the SETI Institute's Project Phoenix (the institute's principal initiative to scan nearby stars for artificially generated radio signals, indicating the presence of technologically advanced extraterres-trial civilizations), to present the Selfish Biocosm hypothesis to the SETI II session of the congress. My paper was entitled "Assessing the Robust-ness of the Emergence of Intelligence: Testing the Selfish Biocosm Hypothesis."

Dr. Tarter, who has acquired a considerable degree of popular fame as the model for Jody Foster's character in the movie *Contact*, had read my original essay proposing the Selfish Biocosm hypothesis in *Com-plexity* (the journal of the Santa Fe Institute). She told me that while she found my hypothesis engaging, she had grave doubts about whether it was genuinely testable, as any hypothesis must be to qualify as "scien-tific." I was in Rio to try to persuade Dr. Tarter and her colleagues in the SETI research community that my idea qualified as a genuinely scientific, albeit highly speculative, hypothesis.

I began my presentation with an abstract summarizing the Selfish Biocosm hypothesis as an assertion that the anthropic qualities that our universe exhibits can be explained as incidental consequences of a cos-mological replication cycle in which a cosmologically extended biosphere supplies two of the essential elements of self-replication identified by the master logician John von Neumann. Further, I went on to say, the hypothesis asserts that the emergence of life and intelligence are key thresholds in the cosmological replication cycle, strongly favored by the physical laws and constants of inanimate nature. Finally, I pointed out in the abstract, a falsifiable implication of the hypothesis is that the emergence of increasingly intelligent life is a robust phenomenon, strongly favored by the natural processes of biological evolution. At this point, I checked my audience of prominent SETI researchers from all over the world for facial expressions indicating disdain or incredulity. I was delighted to see that they seemed to be listening intently and continued my presentation.

scientific field in the broad portfolio of disciplines comprising the astrobiology research endeavor.

After summarizing Lee Smolin's daring cosmological conjecture and its unfortunate flaws, I summarized the Selfish Biocosm hypothesis as an attempt to remedy the most serious of those defects: the absence of any proposed device or process endowing a replicating cosmos with anything resembling heredity or memory.

Then I got into the meat of the presentation. I began by admitting that in order to qualify as a genuine scientific theory rather than a metaphysical speculation, the Selfish Biocosm hypothesis must be empirically testable. A key implication of the hypothesis, I pointed out, is that the process of progression of the cosmos through critical epigenetic thresholds in its life cycle, while perhaps not strictly inevitable, is relatively robust. One such critical threshold, I noted, is the emergence of human-level and higher intelligence, which is essential to the eventual scaling up of biological and technological processes to the stage at which those processes could conceivably exert an influence on the global state of the cosmos.

The conventional wisdom among evolutionary theorists is that the abstract probability of the emergence of anything like human intelligence through the natural process of biological evolution was vanishingly small. According to this viewpoint, the emergence of human-level intelligence was a staggeringly improbable contingent event. However, I reminded the audience, a few distinguished contrarians like science writer Robert Wright, Harvard biologist Edward O. Wilson and Nobelist Christian de Duve take an opposing position, arguing on the basis of the pervasive phenomenon of convergent evolution and other evidence that the appearance of human-level intelligence was highly probable, if not virtually inevitable.

I then posed a rhetorical question to the researchers: Would it be possible to devise experiments to test these competing theories of the robustness of the emergence of human-level (and higher) intelligence? If so, the results would tend to either falsify or support the Selfish Biocosm hypothesis. I described four possible tests of the hypothesis.

1. **SETI Predictions.** The ongoing SETI research endeavor has so far yielded no positive results. The Selfish Biocosm hypothesis, in conjunction with the principle of mediocrity, implies that the SETI project will eventually succeed. An ongoing pattern of failure would tend to disprove the hypothesis. A demonstrated success would be consistent with it.

2. **Convergent Animal Evolution Toward Sentience in Nonprimate Species.** The proof of convergent evolution toward sentience in nonprimate species would tend to support the Selfish Biocosm hypothesis by indicating that the emergence of intelligence is a relatively robust, species-neutral phenomenon. The ongoing research by the SETI Institute into natural dolphin whistle languages provides a logical starting point for the systematic study of this phenomenon.

3. **Artificial Life Evolution.** Agent-based artificial life and artificial society software offers an opportunity to create, *in silica* (that is to say, inside computer memories), fitness landscapes populated with coevolving artificial life-forms. (See sidebar, page 133.) It may be theoretically possible to coax such artificial organisms into an evolutionary process that could eventually yield what Nobel laureate Gerald Edelman called

a "conscious artifact."[1] Roger Penrose, a distinguished quantum theorist and colleague of Stephen Hawking, contends that this is not possible.[2] The Selfish Biocosm hypothesis and its implication of the robustness of the emergence of human-level intelligence would tend to indicate that Edelman is correct and Penrose is mistaken.

4. **Emergence of Transhuman Intelligence.** It may be possible to develop experimental strategies for exploring the possibility that the process of emergence of ever higher levels of intelligence on Earth is ongoing (and perhaps accelerating), either through the emergence of transhuman machine intelligence or through the aggregation of emerging human/machine communities into one or more global superorganisms. The Selfish Biocosm hypothesis implies the emergence of transhuman intelligence as a necessary condition precedent to the task of cosmological engineering by means of which the hypothesized process of life-assisted cosmological replication takes place.

I wrapped up my presentation by reminding the audience of distinguished scientists that the essence of the Selfish Biocosm perspective is that the universe we inhabit is literally in the process of coming to life. Under this theory, the emergence of life and intelligence are not meaningless accidents in a hostile, largely lifeless cosmos but at the very heart of the vast machinery of creation, cosmological evolution, and cosmic replication. Under the theory, I told Dr. Tarter and her colleagues, the oddly life-friendly suite of laws and physical constants that prevail in our particular universe serve a function precisely analogous to that of DNA in earthly creatures: they furnish a recipe for the ontogenetic development of the mature organism and a blueprint that provides the plan for construction of offspring.

Finally, the hypothesis implies that the phenomenon that we Earth-bound humans perceive as a process of biological evolution is more properly viewed as a subroutine in a vast ontogenetic process by means of which the biocosm grows to maturity and prepares itself for replication. A corollary is that the capacity for biological evolution and emergence is essentially front-loaded into the suite of physical laws and constants that govern the evolution of the cosmos over time in precisely the same manner that the recipe for constructing a mature biological organism is front-loaded into that organism's DNA. Put differently, I concluded, the Selfish Biocosm hypothesis adds to the list of ontological possibilities traditionally contemplated by cosmologists—that the universe may be a great equation, a great computation, or a great accident—the following possibility: that the cosmos may be quintessentially a great unfolding life.

Following the presentation I steeled myself for a skeptical or even hostile reaction to my radical new idea. However, I was not prepared for what happened next. Tom Pierson, who heads the SETI Institute and who was a close colleague of the late Carl Sagan, paused for about ten seconds after the conclusion of my speech and then remarked that he had no questions but only a comment. "I only wish," said Mr. Pierson, "that Carl Sagan could have lived to hear your presentation."

I was dumbstruck by the comment and in a bit of a daze as I left the lecture hall at the RioCentro Convention Center in Rio de Janeiro. Perhaps, I remarked to my wife, Lynda, it was merely that the intellectually adventurous SETI research community is predisposed to thinking outside the box. Perhaps a group of more traditional scientists would have dismissed my hypothesis as rampant speculation bordering on science fiction. But then again, I reminded myself, that had been the first reaction to such previously radical ideas as continental drift and the bacterial origin of mitochondria. Maybe there was hope, after all, that my heretical ideas would be taken seriously by the scientific community.

Months later I learned that my Rio paper had been selected from among the hundreds presented at the Fifty-first International Astronautical Congress to be published in *Acta Astronautica*, the scientific journal of the Paris-based International Academy of Astronautics.[3] Following its publication, I began to receive extraordinarily interesting correspondence from a number of distinguished scientists from all over the world about the Selfish Biocosm hypothesis. Some of the communications were dismissive or highly skeptical. One prominent researcher, for instance, expressed doubt that a highly advanced civilization would choose to "play God" by creating a new cosmos, even if it acquired the technological capacity to do so. However, a majority of the correspondents expressed both strong interest and encouragement. Typical of the favorable comments was the statement by a Princeton cosmologist that "your picture makes a lot more sense than [Lee] Smolin's." It was the most auspicious reception I could have reasonably anticipated for my radical new theory that the entire process of cosmological evolution can plausibly be viewed as a quickening life and an awakening intelligence.

Chapter Twelve

A Second Test: Probing the Eschaton at the End of Time

On the plane ride back from Rio de Janeiro, buoyed by the favorable reception given my paper by the SETI research community, I began to think about a second possible test of the Selfish Biocosm hypothesis. If the hypothesis implied that the emergence of ever greater levels of intelligence was a robust phenomenon, strongly favored by the physical laws and constants of inanimate nature, then did it not also imply that the terminal state of the cosmos yielded by that process would exhibit maximal computational capacity? In other words, the hypothesis seemed to imply that biological processes (and the processes of cosmological engineering of which our descendants may eventually be capable) will, in the distant future, scale up in power and sophistication to the point at which they could exert a global influence on the physical state of the entire cosmos. The hypothesis also seemed to imply that those processes must eventually be capable of achieving maximal computational power in order to (1) replicate the "cosmic code" (the laws and constants of nature that prevail in our universe and that are functionally equivalent in the proposed scenario to the genetic code in an earthly organism) and (2) reliably transmit that cosmic code to a new baby universe.

Here was a falsifiable prediction of the hypothesis with serious destructive potential. (When it comes to scientific hypotheses, this is a very good thing.) If there were *no* plausible final state of the cosmos that was both consistent with the best astrophysical evidence currently available *and* that would possess the physical capacity to achieve maximal computational potential, then the Selfish Biocosm hypothesis would be demolished.

The initial prognosis did not look favorable for the survival of my hypothesis. The imaginative Princeton astrophysicist Freeman Dyson and others had long ago concluded that any long-term cosmological scenario in which the universe "ends in fire" (i.e., in a Big Crunch) precludes the possibility of the persistence of information processing (or life, for that matter) through the Big Crunch era. However, Dyson held out hope that the fate of life and information processing in an "open" (i.e., ever expanding) universe was not so dismal. In a landmark paper published in 1979, Dyson suggested that in such a universe a sufficiently advanced civilization could adopt a strategy of intermittent hibernation in order to both continue information processing indefinitely and rid itself of waste heat generated by the process of computation.[1] (Such waste heat is analogous to the temperature increase generated by the operation of a desktop computer, which is dissipated into the atmosphere with the help of a small cooling fan.)

However, the recent discovery that the expansion of the universe is actually accelerating seemed to jeopardize Dyson's hopeful vision of the future prospects of life and intelligence. As physicists Lawrence M. Krauss and Glenn D. Starkman wrote in 1999 in a particularly gloomy scientific paper entitled "Life, The Universe, and Nothing: Life and Death in an Ever-Expanding Universe,"

> In an eternally-expanding universe life might, at least in principle, endure forever. While global warming, nuclear war and asteroid impacts may currently threaten human civilization, one may hope that humanity will overcome these threats, expand into the Universe, and perhaps even encounter other intelligent life-forms. In any case, if intelligent life is ubiquitous in the Universe, it is reasonable to expect that no local threats can ever wipe the slate entirely clean.[2]

That is the good news. The bad news is that there may be insurmountable barriers to the persistence of conscious life in our universe. As Krauss and Starkman describe those barriers, "We find that the future is particularly discouraging if we live in a comological-constant-dominated universe. In this case, very soon, on a cosmic time-scale, our ability to gather information on the large scale structure of the universe will begin to forever decrease. The decreasing information base in the observable universe is associated with a finite and decreasing supply of accessible energy."[3]

And it is not only in a universe where the expansion rate is accelerating that life and consciousness are eventually doomed to extinction, according to Krauss and Starkman. They assert: "Life's long-term prospects are only slightly less dismal in any other cosmology, however. We argue that the total energy that any civilization can ever recover and metabolize is finite, as is the recoverable information content, independent of the geometry or expansion history of the universe."[4]

In this environment, life and intelligence will eventually be forced into an unappealing choice: indulge in profligate consumption of energy and resources to fuel ever faster computation or slip into an increasingly comatose state in which computation proceeds ever more slowly and eventually grinds to a halt. According to Krauss and Starkman, the options are to "live for the moment in high-powered luxury, or progressively reduce the information theoretic complexity of life until it loses consciousness forever."[5] The only hope held out by Krauss and Starkman is that eventually "strong gravitational effects on the geometry or topology of the universe might effectively allow life, or information, to propagate across apparent causal boundaries, or otherwise obviate the global spatial constraints we claim here."[6]

And precisely what might those strong gravitational effects permit life and intelligence to accomplish? Nothing less than the creation of baby universes! As the two scientists put it, "For example, it might one day be possible to manipulate such effects to artificially create baby universes via wormholes or black hole formation or via the collision of monopoles. Then one might hope that in such baby universes conscious life could eventually appear, or that one might be able to move an arbitrarily large amount of information into or out of small or distant regions of the universe."[7] While conceding that these were "interesting possibilities," Krauss and Starkman nonetheless dismissed them as "vastly more speculative than the other possibilities we have discussed here."[8]

Perhaps so, but at least these possibilities fanned a tiny flame of hope that there might exist a plausible cosmic eschatological scenario consistent with the implications of the Selfish Biocosm hypothesis. That hypothesis is, after all, a proposed narrative about the nature of a replication process that yields baby universes. As two theoretical physicists, Katherine Freese and William H. Kinney, were to subsequently suggest in a 2002 paper entitled "The Ultimate Fate of Life in an Accelerating Universe,"

> There are certainly limitations to Dyson's premise. One alternative cosmology which would violate Dyson's premise would be if the universe oscillates or is cyclic. Then the current accelerating phase might be followed by a subsequent recontraction and then again an expansion, and life could begin all over again. Of course the new burst of life might not have any memory of our current cycle, so that this does not provide an altogether satisfactory solution to the problem of enabling life to continue indefinitely. . . . Another alternative would be to create a universe in a lab, along the lines of suggestions made by Guth and Farhi, and then move into it.[9]

After returning home from Rio de Janeiro, I began to wonder whether it might be worthwhile to revisit a crucial assumption made by Freeman Dyson: that life and information processing could not possibly survive a cosmological scenario pursuant to which the universe begins to recontract and eventually ends in either a Big Crunch or a Big Bounce, the theoretical point of transition between a Big Crunch and a new Big Bang. Was this conclusion necessarily true, I wondered? Or was it theoretically possible that a Big Crunch/Big Bounce scenario was, contrary to Dyson's assumption, physically consistent with the possibility of maximally evolved intelligence and technological sophistication? This was the issue to which I decided to turn my attention.

What I was seeking was a plausible cosmological scenario that would yield not merely the indefinite persistence of life and intelligence, but something far more dramatic: a final state of the cosmos that would be physically consistent with a capacity for maximal computational capability. I decided to provisionally call that hypothesized final cosmic state the "eschaton." The term *eschaton* was a neologism I selected because of its relationship to the word *eschatology*, a theological doctrine dealing with the topic of last or final things. The eschaton, I concluded, would be the hypothetical entity capable of functioning as a von Neumann duplicator in the context of the Selfish Biocosm hypothesis.

The next question was whether there existed a plausible cosmological eschatology that could yield an eschaton with maximal computational capability. A preliminary issue was whether there were candidate physical eschatology scenarios available that would support such a capability. Of particular interest to me was a suggested modification and simplification of the recently proposed "ekpyrotic" cosmological scenario, which provides a framework for rigorously modeling a cyclic universe.

The term *ekpyrotic* was coined by a quartet of scientists from Cambridge, Princeton, and the University of Pennsylvania to describe a radically new cosmological scenario they put forward as an alternative to the canonical theory (inflation) to explain three key puzzles about the current structure and makeup of our universe:

- The flatness puzzle (i.e., why is the observable universe so close to being spatially flat?)

- The homogeneity puzzle (i.e., why are regions of the universe that are causally disconnected from one another so similar?)

- The inhomogeneity puzzle (i.e., what is the origin of the original density fluctuations that are responsible for tiny variations in the nearly uniform cosmic microwave background radiation and that presumably "seeded" the formation of the large-scale structure of the universe [galaxies and groups of galaxies] that we observe today?)

The canonical theory of inflation, which is favored by most mainstream cosmologists, does a good job of explaining these puzzles but at a steep price: it appears incapable of being linked with modern theoretical approaches to quantum gravity such as M-theory.[10] In place of superluminal inflation, the ekpyrotic approach invokes new hypothesized physical phenomena that arise in M-theory, which is based on the assumption that there are physical dimensions in addition to the three visible spatial dimensions.

The proposed "ekpyrotic universe" scenario is so named because of its resemblance to a particular vision of the cosmos articulated by the ancient Greeks. As described by its proponents, the term *ekpyrotic* is "drawn from the Stoic model of cosmic evolution in which the universe is consumed by fire at regular intervals and reconstituted out of this fire, a conflagration called ekpyrosis."[11] The scenario proposes that "the universe as we know it is made (and, perhaps, has been remade) through a conflagration ignited by collisions between branes along a hidden fifth dimension."[12] ("Branes" are theoretical structures in so-called brane-universe theories based on M-theory that assume the existence of extra dimensions; cosmic branes correspond roughly to parallel universes with three or more extended spatial dimensions separated by arbitrarily varying distances in a hypothesized fifth spatial dimension, which is hidden forever from our view.) Under the ekpyrotic scenario, the hot Big Bang and the creation of our visible universe result from the collision of two massive branes moving toward one another in an unseen fifth dimension. The collision releases enormous kinetic energy that is converted into a hot, thermal bath of radiation and matter on the visible brane.

Soon after the ekpyrotic scenario was proposed, its proponents (who include Neil Turok, a colleague of Stephen Hawking at the University of Cambridge, as well as other prominent astrophysicists) put forward a simplified and more elegant variation called the "modified ekpyrotic cyclic universe scenario." In this modified model "the Universe undergoes an endless sequence of cosmic epochs which begin with the Universe expanding from a 'big bang' and end with the Universe contracting to a 'big crunch.'"[13] The "linchpin of the cyclic picture is safe passage through the cosmic singularity, the transition from the big crunch to big bang."[14] M-theory provides the basis for the hypothesized safe passage by offering the possibility that "what appears to be a big crunch" in four dimensions "actually corresponds to the momentary collapse of an additional fifth dimension."[15] Critics of the new scenario—and their ranks appear to be growing—assert that the ekpyrotic cyclic scenario is fatally flawed, in part because M-theory does not, in their view, provide a means for safe passage through the final cosmic singularity.[16]

Why is this arcane issue important? Because in ordinary four-dimensional general relativity theory, the "big crunch is interpreted as the collapse and disappearance of four-dimensional space-time," and there is no indication that a transition is possible between a Big Crunch and a new Big Bang.[17] By contrast, under the modified ekpyrotic cyclic universe scenario "temperature and density are finite as one approaches the crunch," and "there is nothing to suggest that time comes to an end when the fifth spatial dimension collapses"; on the contrary, the "most natural possibility is that time continues smoothly" across the Big Crunch/Big Bang threshold.[18] While the characteristics of the Big Crunch/Big Bang transition have not yet been modeled rigorously in M-theory, it was sufficient for my purposes that the ekpyrotic cyclic universe scenario represented a plausible cosmological eschatology, derived from M-theory and consistent with the latest observational data. Under this theory, the eschaton would correspond to the physical state of the cosmos immediately prior to and during the Big Crunch.

A major caveat is in order at this point. It must be conceded that the modified ekpyrotic cyclic universe scenario is a relatively new conceptual entrant in the rapidly inflating field of string-theory-inspired cosmological conjecture and that it has already drawn heated and intensifying criticism from such prominent astrophysicists as Andrei Linde.[19] However, the scenario has already demonstrated (at least in the judgment of its proponents) that it possesses key advantages over alternative scenarios, including its incorporation of the currently observed phenomenon of cosmic acceleration. As noted by its proponents,

> What is the role of dark energy [which is another name for the mysterious antigravitational force termed "lambda" in chapter 2] and the current cosmic acceleration [in the modified ekpyrotic cyclic universe scenario]? Clearly, dark energy and the current cosmic acceleration play an essential role in the cyclic model both by reducing the entropy and black hole density of the previous cycle, and triggering the turnaround from an expanding to a contracting phase. (In all other cosmologies to date, including inflation, dark energy has no essential role.)[20]

The modified ekpyrotic cyclic universe scenario provided a promising new cosmic eschatological model that might furnish a physical template for the hypothesized process of cosmological replication that is the heart of the Selfish Biocosm hypothesis. To be clear, my goal was not to pass judgment on the validity or lack thereof of the newly proposed cyclic scenario or to assess its merits vis-à-vis the theory of eternal chaotic inflation pioneered by Linde. My objective was far narrower: to assess the consilience between that scenario and my hypothesis. This was, I felt, an interesting way of determining how snugly my Selfish Biocosm hypothesis might fit with at least one adventuresome new cosmological model.

The crucial question I next faced was whether—and, in principle, how capably—the hypothesized final state of the cosmos yielded by the modified ekpyrotic cyclic universe scenario could, at least in theory, compute. As previously noted, the canonical assumption has been that any scenario pursuant to which the universe "ends in fire" (i.e., a Big Crunch) precludes the possibility of the persistence of computation through the Big Crunch era. While the modified ekpyrotic cyclic universe scenario sketched out above casts some doubt on this conventional wisdom in and of itself (by eliminating the necessity for a four-dimensional singularity at the instant of the Big Crunch), the most powerful indication that this scenario requires a rethinking of the canonical assumption came from a branch of science far removed from cosmology: computational theory.

In a 2000 issue of *Nature*, MIT computer scientist and complexity theorist Seth Lloyd sought to explore the ultimate physical limits of computation as constrained by three fundamental constants of nature: the speed of light, the quantum scale, and the gravitational constant. The "ultimate computer" described by Lloyd's exercise would be a computing device as powerful as the laws of physics allow. The "typical state" of the memory of Lloyd's hypothesized computing device looks, according to Lloyd, "like a plasma at a billion degrees Kelvin—like a thermonuclear explosion or a little piece of the Big Bang." This was because "to take full advantage of the memory space available" the ultimate computing device "must turn all its matter into energy."[21]

I had hit pay dirt. If Lloyd's conclusions were correct, they indicated that the characteristics of the eschaton (i.e., the final state of the cosmos just prior to and during the "Big Bounce," which is the theoretical point of transition from the contraction phase of the universe [the Big Crunch] to a new expansion phase [a new Big Bang]) under the modified ekpyrotic cyclic universe scenario—a nonsingular condition characterized by extremely high temperature and density—are precisely congruent with the predicted characteristics of an ultimately powerful computational device. In short, under this scenario the eschaton could plausibly constitute the most powerful computer permitted to exist by the laws of physics.

Now I was ready to publish a new paper setting forth a second major falsifiable implication of the Selfish Biocosm hypothesis: that there exists a plausible final state of the cosmos that exhibits maximal computational potential. This predicted final state—the eschaton—appeared under the modified ekpyrotic cyclic universe scenario to be consistent with Lloyd's description of the physical attributes of the ultimate computational device: a computer as powerful as the laws of physics will allow.

There was just one small problem: it was going to be difficult to find a top-flight peer-reviewed scientific journal with a sufficiently broad focus to encompass both cosmology and computational theory and headed by a sufficiently bold editor willing to publish a highly speculative paper that drew upon these two fields. Then I remembered a sample copy of a journal I had stumbled across at the International Astronautical Congress in Rio de Janeiro. The *Journal of the British Interplanetary Society* (known in the space-science community as *JBIS*) is the oldest continuously published journal in the world devoted to spaceflight and related topics. The storied accomplishments of *JBIS* include publication of early speculations about the feasibility of Earth-orbiting communications satellites as well as papers on such unconventional topics as interstellar travel and colonization. This futuristic journal became my first choice among potential publication venues.

I am happy to report that my new paper, which I titled "Assessing the Computational Potential of the Eschaton: Testing the Selfish Biocosm Hypothesis," was accepted and published in 2002 in *JBIS*.[22] (A copy of the paper as published in *JBIS* is included in appendix 2.) The reaction in the scientific community has been similar to that generated by my previous Selfish Biocosm papers: considerable interest tempered by great skepticism. (For instance, one prominent astrophysicist, from the Institute for Advanced Studies in Princeton, asked me whether we would have to build an eschaton in order to validate my theory. I responded to him that I hoped not because, if so, validation would have to wait many billions of years until the experimental setup would be ready!)

The skepticism results, at least in part, from the very idea that life and intelligence could, in the far distant future, evolve to a point of such power and sophistication that it could exert a global effect on the entire physical cosmos and indeed enable the cosmos to reproduce. It may be that such an end point in the evolutionary process is, as Darwin opined, far beyond the capacity of our meager monkey brains to comprehend or even imagine. But perhaps not. Maybe we can attempt a rudimentary speculation about what life and thought would be like at this late stage in the evolution of the universe.

To evolve the kind of information-processing capacity that the hypothesized eschaton possesses would be to progress toward a state that might be described as one of computational rapture. As this state is approached, the tempo and volume of computation would likely accelerate exponentially. But there would probably be no subjective sense of haste or urgency on the part of whatever mind or community of minds might then be alive because time itself would have slowed to a crawl as a result of relativistic effects. Even though from our perspective only a few microseconds might remain to accomplish what is arguably the most important task in the long history of the universe—arranging for its successful reproduction—the entities responsible for this task would perceive that a virtually endless corridor of time stretched before them.

The speculative physicist Frank Tipler described a similar Big Crunch scenario:

> [A]lmost all closed universes undergo "shear" when they recollapse, which means they contract at different rates in different directions. . . . [T]his shearing gives rise to a radiation temperature difference in different directions, and this temperature difference can provide sufficient free energy for an infinite amount of information processing between now and the final singularity, even though there is only a *finite* amount of proper time between now and the end of time in a closed universe. Thus, although a closed universe exists for only a finite proper time, it nevertheless could exist for an infinite subjective time, which is the measure of time that is significant for living beings.[23]

If Tipler is correct—and his views certainly seem to coincide with standard interpretations of Einstein's theory of relativity, which predicts that the dimension of time will expand toward infinity as the final singularity is approached—then the instant of the eschaton will be perhaps the closest that mortals can come to experiencing eternity—the oceanic sense of a vast expanse of future time, extending without perceptible limit to the temporal horizon of the universe. From the subjective viewpoint of the minds inhabiting this unique setting, the eschaton will be an utterly spacious temporal environment; to paraphrase the poet Andrew Marvell, it will offer world enough and time within which mind can finally achieve the glorious consummation that was the promise and hard-won legacy of eons upon eons of prior evolution.

In his speech accepting the 2000 Templeton Prize for Progress in Religion, Freeman Dyson had this to say about the nature of mind in an evolving universe:

> It appears that mind, as manifested by the capacity to make choices, is to some extent inherent in every atom. The universe as a whole is also weird, with laws of nature that make it hospitable to the growth of mind. I do not make any distinction between mind and God. God is what mind becomes when it has passed beyond the scale of our comprehension. God may be either a world-soul or a collection of world-souls. So I am thinking that atoms and humans and God may have minds that differ in degree but not in kind. We stand, in a manner of speaking, midway between the unpredictability of atoms and the unpredictability of God. Atoms are small pieces of our mental apparatus, and we are small pieces of God's mental apparatus.[24]

Dyson's speech provides us with an awe-inspiring way to think about humanity's role in the unfolding cosmic drama leading up to the moment of the eschaton: a small organ in the evolving mind of the universe and a primitive ancestor of the transcendently capable entities our distant progeny may one day become.

Is it reasonable to equate such a supremely advanced entity or community of entities to the concept of deity, as Dyson did? An intriguing answer was offered by that professional skeptic Michael Shermer, publisher of *Skeptic* magazine and author of the "Skeptic" column for *Scientific American.* Sounding, for once, rather more credulous than skeptical, Shermer had this to say about the principle of equivalence articulated by Dyson (i.e., that God is what mind becomes when it has passed beyond the scale of our comprehension):

> [An observation by science fiction author Arthur C. Clarke that "any sufficiently advanced technology is indistinguishable from magic"] stimulated me to think about the impact the discovery of an extrater-

restrial intelligence (ETI) would have on science and religion. To that end, I would like to immodestly propose Shermer's Last Law . . . : "Any sufficiently advanced ETI is indistinguishable from God." God is typically described by Western religions as omniscient and omnipotent. Because we are far from possessing these traits, how can we possibly distinguish a God who has them absolutely from an ETI who merely has them copiously relative to us? We can't. But if God were only relatively more knowing and powerful than we are, then by definition the deity *would* be an ETI![25]

Part Five

Intimations of Cosmic Grandeur

Part 5 of the book speculates about the methods that a highly evolved superintellect might employ, in the far distant future, to accomplish the daunting feat of universe propagation. Specifically, part 5 considers how such a supermind might create baby universes containing the full suite of life-friendly physical laws and constants that prevail in our cosmos, thus replicating the "genotype" of our particular universe. This section also puts forward the notion that if the space-time continuum (i.e., our cosmos in its entirety) constitutes a closed loop connecting one doorway of time (the Big Bang) to another (the Big Crunch), then our oddly life-friendly universe could conceivably, in the words of Princeton astrophysicist J. Richard Gott III, be capable of the ultimate act of self-replication: it could serve as its own mother.

BIOCOSM

Chapter Thirteen

A Cosmological Origin of Biological Information

The Selfish Biocosm hypothesis claims that life and highly evolved intelligence are central players in the drama of cosmological evolution, responsible for both guiding the process (at least in its later stages) and accomplishing the daunting task of cosmic replication. As we have seen, several falsifiable implications of the hypothesis concern the predicted state of power and sophistication that intelligence will need to attain in future cosmic ages to satisfy the exacting requirements of the theory.

Another set of falsifiable implications generated by the hypothesis concern not the future, but rather the distant past—specifically the pre–Big Bang cosmological era. It used to be considered scientifically unacceptable to ask what happened before the Big Bang. That singular event was thought to have launched not only the history of the universe we inhabit but also to have commenced the very flow of time itself. Asking what happened *before* time began was considered logically incoherent—like asking which point on a perfect circle is the beginning and which point is the end, or what is north of Earth's North Pole.

Then the assumption that it was fruitless to investigate the pre–Big Bang era began to be challenged. An important step was taken when Russian-born physicist Andrei Linde articulated the notion of an eternally self-reproducing universe replicating itself chaotically into an infinite number of exponentially enlarged baby universes with different laws of low-energy physics. In Linde's brilliant vision there exists a vast plenitude of universes that both predate and postdate our own—a staggeringly immense "multiverse" ensemble in which new Big Bangs are going off all the time and will continue to explode, in exponentially increasing numbers, for all of eternity. With the concept of this scenario, it was not only logical but absolutely imperative to ask what might have happened before our particular Big Bang occurred.

While Linde's theoretical approach of eternal chaotic inflation broke the taboo against asking what transpired before the Big Bang, it did not shed light on one of the central issues raised by the Selfish Biocosm hypothesis: How did our particular cosmos get to be so life-friendly? Linde's explanation for this curious fact centered on a classic statement of the weak anthropic principle. As he put it with commendable candor in a scientific paper delivered at Stephen Hawking's sixtieth birthday conference in Cambridge,

> The situation changed completely when it was found that eternal inflation occurs in the chaotic inflation scenario. . . . In such a case the universe can probe all possible vacuum states of the theory. This for the first time provided physical justification of the anthropic principle. Indeed, if this scenario is correct, then physics alone cannot provide a complete explanation of the universe. The same physical theory may yield large parts of the universe that have diverse properties. According to this scenario, we find ourselves inside a four-dimensional domain with our kind of physical laws not because domains with different dimensionality and with alternate properties are impossible or improbable, but simply because our kind of life cannot exist in other domains.[1]

It shows no disrespect to the brilliant Dr. Linde to respond that the weak anthropic principle does not provide an explanatory paradigm that is entirely intellectually satisfying.

Cosmic Replication in the Distant Past

The Selfish Biocosm hypothesis suggests that highly evolved life and intelligence play a central role in the process of cosmic replication. Moreover, the principle of mediocrity (i.e., the notion that there is nothing special or unique about the position of our particular universe in the cosmic replication sequence) suggests that the current "cosmic generation" represents neither the first nor the last iteration of the replication cycle. This leads to a new set of possible falsifiable implications of the hypothesis. We have previously considered a few of the implications of the theory for the distant future. What might the hypothesis imply with respect to the distant past?

Related topics have been seriously considered by only a handful of scientists. The late British astronomer Fred Hoyle, a self-proclaimed atheist famous for the now discredited theory of a steady-state universe but less well known for his astonishing predictions concerning the anthropic aspects of stellar nucleosynthesis inside giant supernovae, concluded that the most straightforward explanation for the astonishing array of life-friendly coincidences embedded in the laws and constants of nature was that a superintellect located somewhere in space and time had somehow deliberately engineered the laws of physics to make it possible for carbon-based life and intelligence to evolve. In a 1982 essay he explained this possibility:

> Would you not say to yourself, "Some super-calculating intellect must have designed the properties of the carbon atom, otherwise the chance of my finding such an atom through the blind forces of nature would be utterly minuscule?" Of course you would. A common sense interpretation of the facts suggests that a superintellect has monkeyed with physics, as well as with chemistry and biology, and that there are no blind forces worth speaking about in nature. The numbers one calculates from the facts seem to me so overwhelming as to put this conclusion almost beyond question.[2]

The unknown superintelligence that preceded us, Hoyle believed, put together as a "deliberate act of creation" a universe that was suitable for carbon-based life and the evolution of intelligence.[3] Hoyle stressed that the superintellect of which he was speaking was not a supernatural deity but a natural entity whose essence we could ultimately aspire to understand. Far from being religiously inspired, the idea that such a naturally occurring superintellect might have existed and might have been responsible for the deliberate engineering of the basic laws of nature was, in Hoyle's view, deeply antithetical to the proreligion bias of Western civilization and Western science. He stated, "The idea that the intelligence that designed carbon-based life is squarely within the Universe of normal cause and effect is one that has had an uncomfortable reception in the contemporary western world because in conformity with Judaeo-Christian tradition it seems to be the real wish of western astronomers to invoke supernatural ultimate causes from outside the Universe."[4]

Hoyle proclaimed himself to be unrepentantly Greek in his philosophical orientation in that he subscribed to the rationalist philosophy that there was an ultimate, discoverable order in the universe, in contrast to the view of Western monotheistic religion (and perhaps Western science) that the rational paradigms proffered by scientists can go only so far in explaining the ultimate structure of the cosmos. Hoyle believed that the underlying animus of Western science was the conviction that beneath every equation was ultimately the unfathomable mind of God. Hoyle's view was that we could not only know the mind of the author of nature but that "we are the intelligence that preceded us in its new material representation—or rather, we are the reemergence of that intelligence, the latest embodiment of its struggle for survival."[5]

Another contemporary scientist who has articulated similar ideas is the astronomer Edward Harrison. In an audacious scientific paper published in Britain in 1995 in the prestigious *Quarterly Journal of the Royal Astronomical Society*, Harrison suggested that our universe was created by life-forms possessing superior intelligence existing in another physical universe in which the constants of physics were finely tuned and therefore essentially similar to our own.[6] Further, Harrison suggested that highly intelligent beings, perhaps including our own descendants in the far future, might possess not only the knowledge to design but also the technology to build baby universes. Finally, Harrison conjectured that the very comprehensibility of the universe to the human mind might be a subtle clue that the universe was, in fact, designed by minds basically similar to our own.

Finally, in an unpublished paper entitled "On the Role of Life in the Evolution of the Universe," Kansas State University mathematician Louis Crane suggested that Harrison's speculations required us to consider the possibility that the historical processes of biological and cosmological evolution are inseparably linked:

> In the first place, the origin and evolution of life [can] no longer [be viewed as] a mere accident. Rather it is deliberately coded into the fine tuning of the physical laws. Since the development of life and of the universe are joined into a unified evolutionary process, they can be viewed from the point of view of purpose, just as it makes sense to speak of the purpose of an organ of a developing animal, even though the development of the animal is entirely within the scope of physical law. Secondly, intelligence and its ongoing success are no longer a small and unimportant accident in an enormous universe. Rather they are the precondition for the existence and reproduction of the universe. The world around us was created by something like us, and is structured, as if deliberately, to produce us and nurture us. We have a larger purpose which goes beyond ourselves of sustaining and recreating the universe.[7]

Crane recognized that such speculations invade the province of thought traditionally reserved to religion but offered no apologies for the transgression. He commented, "If this [scenario] is eventually demonstrated to be true, it should have cultural implications beyond the sphere of science itself. The traditional spheres of religion and science will fuse, but on a new basis, and with no mystical illusions."[8]

How Did It All Begin?

The courageous speculations of Harrison, Hoyle, Crane, and a handful of other iconoclastic scientists unfortunately shed no light on a central problem raised by the Selfish Biocosm hypothesis: Who or what initially launched the process of cosmic reproduction? How did life and intelligence first become possible? How did the "first" universe become sentient and thus capable of seeding its progeny?

The tests of the Selfish Biocosm hypothesis previously proposed do not address the issue of how the information governing the process of biological evolution and emergence that is hypothesized to lead up to the cosmological replication event can plausibly arise in the first instance. Yet a falsifiable implication of the hypothesis is that there must exist a natural cosmological mechanism by means of which the specified complexity that is the essence of biological information can initially arise. Put differently, since the hypothesis implies that the origin of life and intelligence is a cosmic imperative, encoded as a subtext to the laws and constants of inanimate nature, it follows as a falsifiable implication of that hypothesis there must be a plausible natural means through which that life-friendly cosmic code was initially composed. But what could that natural process possibly be? Who or what could have first authored the cosmic code?

Directed Panspermia: An Inadmissible Tower of Turtles

Faced with the daunting puzzle of life's origin, Darwin famously opined that this particular challenge, no less than the "extreme difficulty or rather impossibility of conceiving this immense and wonderful universe, including man with his capacity of looking far backwards and far into futurity, as the result of blind chance or necessity,"[9] was perhaps forever beyond the reach of what he called humanity's godlike intellect. Equally intimidated by the origin-of-life puzzle, DNA codiscoverer

Francis Crick sought intellectual solace in the speculative concept of "directed panspermia"—the notion that life might be able to deliberately perpetuate and multiply itself throughout the cosmos by means of a kind of intentional diaspora.[10]

The problem is that this proposed mechanism, while increasing the odds that life could permeate the cosmos, merely postpones the problem of the origin of biological information. Crick's hypothesis offers an exotic yet plausible response to the dilemma posed by physicist Enrico Fermi about the troublesome lack of evidence of extraterrestrial life (Fermi asked, If the origin and evolution of life throughout the universe is chemically predestined, why is it we have acquired no direct evidence of the hypothesized phenomenon?) by asserting that all earthly organisms may, in fact, be the remote progeny of extraterrestrial microbes. Yet the hypothesis does not seek to probe the central mystery of the origin of the primal coding mechanism and thus of the origin of life considered as the origin of specified complexity. In short, directed panspermia is a laudable attempt to explain life's dispersion but not its commencement. Viewed as a proffered solution to the origin puzzle, Crick's hypothesis must be disqualified as what physicist John Wheeler would call an inadmissible "tower of turtles standing one on the other"—a framework of ideas that indulges in infinite regress and thus sidesteps the ultimate issue.[11]

The problem of infinite regress is precisely the same whether the issue is the origin of organic life or the origin of an entire intelligently designed universe governed by a life-friendly cosmic code. Astronomer Edward Harrison characterized the dilemma as follows:

> The creation [of such] a universe . . . requires a high level of intelligence, thus raising the question of how the first universe began. A parallel problem concerns the origin of life on Earth. The probability of life originating on a planet is exceedingly small. No doubt the Galaxy teems with lifeless planets on which the conditions were never favourable for the origin of life. Life originated on Earth because of its unusual conditions. . . . In cosmogenesis, one possibility—a varia-

> tion on the anthropic principle—is that an initial ensemble of universes, in which the fundamental parameters have random variations, contains at least one member in which intelligent life is possible. This member is the "intelligent" mother universe. Thereafter, by reproduction, intelligent universes dominate the ensemble, and the original unintelligent members then form a vanishingly small fraction of the whole. As on Earth, life originates with the first self-reproducing molecule, produced by chance, and thereafter it proliferates and dominates. . . . We are still left with the ultimate question: . . . who created the initial cosmic ensemble in the anthropic principle? . . . Does the prospect of infinite regress mean the discussion has now moved into realms of reality where creative agents exist in universes no longer comprehensible to the human mind?[12]

As we shall see, no such surrender to incomprehensibility is required.

The Ekpyrotic Cyclic Universe as a Closed Timelike Curve

As noted earlier, in a paper published in *JBIS* I suggested that the recently proposed modified ekpyrotic cyclic universe scenario offered a plausible physical template for the process of cosmological emergence and replication that is the heart of the Selfish Biocosm hypothesis. In particular, I suggested that a falsifiable implication of the hypothesis—that there exists a plausible final state of the cosmos that exhibits maximal computational potential—appeared to be consistent with the ekpyrotic cyclic scenario's prediction of the characteristics of the final state of a cyclic cosmos immediately prior to and during a Big Bounce era and also with Seth Lloyd's description of the ultimate computational device: a computer as powerful as the laws of physics will allow.

For purposes of the present inquiry into possible ultimate sources of the cosmic code, it is useful to summarize another predicted characteristic of the ekpyrotic cyclic scenario: the uninterrupted flow of time across the threshold of the Big Bounce era. In this scenario "the Universe undergoes an endless sequence of cosmic epochs which begin with the Universe expanding from a 'big bang' and end with the Universe contracting to a 'big crunch.'"[13] The most important aspect of that scenario for purposes of the present inquiry is that it predicts that time (and thus causation and the flow of information) continues smoothly across the Big Crunch/Big Bang era. This prediction opens up a radically novel potential venue for the origin of biological information: the distant future.

In 1997, Princeton astrophyscists J. Richard Gott III and Li-Xin Li posed this intriguing question: Could our universe conceivably have spawned itself?[14] Beginning with the recognition that "a remarkable property of [Einstein's theory of] general relativity is that it allows solutions that have closed timelike curves (CTCs)"[15]—hypothetical configurations of space and time where gravity is sufficiently strong to bend the space-time continuum into a looping configuration that allows future events to influence the past—the two scientists pointed out that, absent some rule like the chronology protection conjecture proposed by Stephen Hawking (which states that the laws of physics conspire to forbid the actual manifestation of CTCs, at least at the macroscopic scale), the "Universe can be its own mother."[16] Under the CTC cosmological scenario "the Universe neither tunneled from nothing, nor arose from a singularity; it created itself."[17]

Commenting on the potential relationship of this CTC scenario to the conjecture by Edward Harrison that our life-friendly universe might be the artifact of a prior advanced civilization—a baby universe created "in the lab" by some supercivilization in a prior universe—Gott and Li noted that Harrison was able to explain multiple generations of artificially created baby universes by this mechanism, except for the first one. As Gott and Li put it, "This seems to be an unfortunate gap. In our scenario, suppose that first universe simply turned out to be one of the infinite ones formed later by intelligent civilizations. Then the Universe—note capital U—would be multiply connected, and would have a region of CTCs; all of the individual universes would owe their birth to some intelligent civilization in particular in this picture."[18]

The ekpyrotic cyclic universe scenario adds to the conjecture put forward by Gott and Li—that the Universe might be its own mother—the crucially important possibility that time (and thus causation and information) might flow smoothly from a Big Crunch era (an epoch characterized, as previously shown, by maximal computational potential) to a Big Bang era and that the two eras might be linked by means of a CTC. This implies that the point of ultimate origin of biological information—of specified complexity—might plausibly lie in our distant future, which would be adjacent to our distant past on the unbounded loop of a CTC.

Causation from the Super-Copernican Perspective

The principal reason that Darwin doubted the capacity of the human mind to probe profound cosmic mysteries like these was his skepticism that "the mind of man, which has, as I fully believe, been developed from a mind as low as that possessed by the lowest animal, [can] be trusted when it draws such grand conclusions."[19] Always the cautious skeptic, Darwin never forgot that, to paraphrase the concluding passage of *The Descent of Man*, mankind's godlike intellect as well as his bodily frame still bore the indelible imprint of humanity's lowly origin.

However, Darwin's own monumental achievements and those of countless other pioneers of human science and civilization provide ample reason to believe that such limitations are not insurmountable obstacles to human comprehension of even the most cosmically vast and counterintuitive physical phenomena.

For purposes of the present inquiry, the key perspective is offered by what physicist John Wheeler calls the super-Copernican principle. Derived from the Copenhagen interpretation of quantum physics, this "principle rejects the now-centeredness of any account of existence as firmly as Copernicus rejected here-centeredness."[20] According to this principle, the future can have at least as important a role in shaping the present moment as the past. The most important aspect of Wheeler's insight is not that we must embrace the specific mechanism of retroactive causation favored by Wheeler and the advocates of the Copenhagen interpretation (the retroactive impact on quantum phenomena of observer-participancy), but rather that we should be open to counter-intuitive notions of causation, if they appear to be consistent with novel yet mathematically plausible accounts of physical reality.

In particular, the vision of the cosmos as a closed timelike curve offers a new paradigm that may allow us to grasp new theoretical possibilities for the origin and nature of biological information and the specified complexity it exhibits. According to this paradigm, the process of biological information generation can be viewed as an essentially eternal wave function continuously yielding ontogenetic complexification. This wave function, which might be provisionally termed the "ontogon"—a neologism derived from the biological term *ontogeny*—moves along the closed timelike curve from what we call the past to what we call the future and back again across the Big Crunch era to a new Big Bang era without disruption by a final singularity.

Causation defines the relationship between all points on the CTC, but the relationship of cause and effect is not temporally restricted in the sense we naively perceive. As Wheeler put it with uncanny prescience (though with a different causal mechanism in mind), the history of the cosmos "is not a history as we usually conceive history. It is not one thing happening after another after another. It is a totality in which what happens 'now' gives reality to what happened 'then,' perhaps even determines what happened then."[21] Because the CTC is curved *and* timelike *and* closed *and* unblemished by a final singularity, each point on the CTC is both the cause *and* effect of every other point.

The CTC that is hypothesized to be our cosmos thus may be a classic autocatalytic set, what Wheeler ventured to call a "self-excited circuit"[22] and a "grand synthesis, putting itself together all the time as a whole."[23] The implication for the origin of biological information should be apparent: not only the universe but also life itself (and the specified complexity it embodies) can be its own mother under this scenario.

Speculating on this possibility (but without the benefit of the specific scenario provided by Gott, Li, and the modified ekpyrotic cyclic universe proposal), the astrophysicist Paul Davies had this to say in an online *Edge* interview:

QUESTION: You mention aliens. Who are the aliens?

DAVIES: We don't know. We could be totally alone in the universe; at this particular time it's impossible to say. But we can speculate that there might be life, even intelligent life, elsewhere.

QUESTION: Could they be our ancestors? Or our God?

ANSWER: Descendants maybe, not ancestors. Well, I guess if it's possible to travel through time as well as through space, we can imagine the universe being populated by a single species far into the future and also backwards into the past, so they could also be our ancestors too. It wouldn't be necessary to have life popping up independently in many different places. That would be a curious twist on the time-travel story. We could go backwards in time and seed other planets with life at an earlier epoch. Yes, that's always conceivable.[24]

Whether this possibility is not only conceivable but also falsifiable in a Popperian sense (thus qualifying as a scientific speculation) is far from clear. However, at the very least, the possibility is scientifically plausible.

Chapter Fourteen

The Tool Kit of Cosmic Ontogeny and Reproduction

Astrobiology at NASA

Astrobiology is the study of life in the universe. As Peter Ward and Donald Brownlee point out in their book Rare Earth, *the field actually encompasses a wide range of converging scientific disciplines: "What makes this [Astrobiology Revolution] so startling is that it is happening not within a given discipline of science, such as biology in the 1950s or geology in the 1960s, but as a convergence of widely different scientific disciplines: astronomy, biology, paleontology, oceanography, microbiology, geology, and genetics, among others.*[9]

The astrobiology research initiative at NASA seeks to answer

(continued on next page)

Detailed conjecture concerning the precise technology by means of which highly evolved life-forms could actually copy the life-generating cosmic code and then transmit it to a new baby universe is beyond the scope of current science. However, it is possible to preview at least some of the issues that would arise if one were to contemplate the design of a tool kit of cosmic ontogeny and reproduction.

M-Theory and the Possibility of Calabi-Yau Shape Engineering

A lecture by Stephen Hawking entitled "Quantum Cosmology, M-Theory and the Anthropic Principle" suggests that a fruitful area of initial speculation might be M-theory and in particular the implications of M-theory with respect to the period of seemingly random evolution of "Calabi–Yau" spaces in the earliest post–Big Bang era.[1] Calabi-Yau spaces are the six-dimensional geometrical shapes hypothesized by M-theory in which compactified spatial dimensions beyond the three familiar extended spatial dimensions are curled up tightly. According to the theory, the oscillation of superstrings within these minute spaces gives rise to a whole menagerie of subatomic particles, including the elusive graviton.

One possibility to explore might be the deliberate engineering of such shapes during a period that some superstring theorists refer to as the "*prehistory* of the universe—starting long before what we have so far been calling time zero—that leads up to the Planckian cosmic embryo."[2]

Could this prehistorical period conceivably coincide with or encompass the eschaton or Omega Point era of a predecessor cosmos and the artifacts of that era, including deliberately engineered Calabi-Yau shapes? If so, can we conceive of mechanisms by means of which such engineered shapes could be transmitted by a sufficiently advanced life-form across the time zero threshold to a new baby universe? Probably not yet.

One startling implication of the possibility of Calabi-Yau shape engineering is that this could be a mechanism not only for transmitting a "genetic code" to a new baby universe but also for sending a kind of message to its future inhabitants of the sort hypothesized by the late Heinz Pagels (as discussed in the first chapter of this book). One scientist who has thought seriously about this possibility is the astrophysicst Andrei Linde, whom we encountered in the context of his theory of eternal chaotic cosmic inflation.

In an audacious scientific paper entitled "The Hard Art of Universe Creation," which was published in 1991, Linde asked whether it might be theoretically possible for the fabricators of a new baby universe created "in the lab" to send a message to future living creatures who might one day inhabit such an artificially created cosmos. The only possibility for accomplishing this feat, Linde concluded, would be to embed the message in the laws of physics that would prevail in the new universe. He theorized:

> It seems that the only way to send a message to whose who will live in the universe we are planning to create is to encrypt it into the properties of the vacuum state of the new universe, i.e., into the laws of the low-energy physics. Hopefully, one may achieve it by choosing a proper combination of temperature, pressure and external fields, which would lead to creation of the universe in a desirable phase state.[3]

In a brilliant insight whose import has not been widely acknowledged, Linde realized that what is normally viewed as a *weakness* of superstring theory or M-theory—that it does not uniquely predict a single set of laws and constants of low-energy physics but is rather ca-

three fundamental questions:

- *How does life begin and develop?*
- *Does life exist elsewhere in the universe?*
- *What is life's future on Earth and beyond?*

NASA has promulgated ten specific science goals in an attempt to answer these profound questions:

- *Goal 1: Understand how life arose on Earth.*
- *Goal 2: Determine the general principles governing the organization of matter into living systems.*
- *Goal 3: Explore how life evolves on the molecular, organism, and ecosystem levels.*
- *Goal 4: Determine how the terrestrial biosphere has coevolved with the planet.*
- *Goal 5: Establish limits for life in environments that provide analogues for conditions on other worlds.*
- *Goal 6: Determine what makes a planet habitable and how common these worlds are in*

the universe.

- *Goal 7: Determine how to recognize the signature of life on other worlds.*
- *Goal 8: Determine whether there is (or once was) life elsewhere in our solar system, particularly on Mars and Europa (the moon of Jupiter thought to contain the largest water ocean in the solar system beneath a solid crust of ice).*
- *Goal 9: Determine how ecosystems respond to environmental changes on timescales relevant to human life on Earth.*
- *Goal 10: Understand the response of terrestrial life to conditions in space or on other planets.*

Proponents of the Selfish Biocosm hypothesis might wish to propose that NASA consider adding an eleventh goal: Determine how to recognize the signature of life and intelligence in the laws of physics and chemistry that prevail in our cosmos!

pable of generating a whole menagerie of seemingly arbitrary sets of such physical rules—is, in fact, a crucial *strength* if one of the possible functions of those physical laws and constants is to transmit information. As Linde puts it,

> The corresponding message can be long and informative enough only if there are extremely many ways of symmetry breaking and/or patterns of compactification in the underlying theory. This is exactly the case, e.g., in the superstring theory, which was considered for a long time as one of the main problems of this theory. Another requirement to the informative message is that it should not be too simple. If, for example, masses of all particles would be equal to each other, all coupling constants would be given by 1, etc., the corresponding message would be too short. Perhaps, one may say quite a lot by creating a universe in a strange vacuum state. . . . The stronger is the symmetry breaking, the more "unnatural" are relations between parameters of the theory after it, the more information the message may contain. Is it the reason why we must work so hard to understand strange features of our beautiful and imperfect world? Does this mean that our universe was created not by a divine designer but by a physicist hacker? If it is true, then our results indicate that he did a very difficult job. Hopefully, he did not make too many mistakes.[4]

The analogy to the key property of DNA that makes it capable of encoding a construction plan for an organism is quite precise. It is the very fact that the *sequence of four nucleotides* (adenosine, cytosine, guanine, and thymine) in a strand of DNA is not prefigured by the inherent chemical properties of the DNA molecule that allows DNA to function as a superb genetic coding mechanism. If the "letters" in the DNA "alphabet" could not be sequenced in an arbitrary order corresponding to the "recipe" for a particular organism, then DNA would be a woefully inadequate vehicle for encoding and transmitting genetic information.

So too, as Linde has demonstrated, it is the inherent flexibility of superstring theory or M-theory—a conceptual framework theoretically capable of generating a whole menagerie of wildly different universes exhibiting an arbitrary mix of disparate physical laws and constants—that makes it theoretically possible for those laws and constants actually manifested in a particular universe to encode a kind of "message," which is the functional counterpart to a message encoded in an earthly organism's DNA.

Stephen Wolfram: Hacking the Software of Everything

A second speculative question concerning the transmission of the cosmic code across the time-zero threshold is this: How much information must be transmitted in order to permit a new baby universe to reconstruct itself in the image of its parent? This issue may have become considerably more tractable as a result of recent work in complexity theory, notably the extraordinary contributions of Stephen Wolfram in his 2002 book, *A New Kind of Science.*[5]

Wolfram is an exceptional scientist by any measure. A certified wunderkind, he published his first scientific paper when he was fifteen years old. By age twenty he had earned a Ph.D. in theoretical physics from the California Institute of Technology. A year later, he won a MacArthur Foundation "genius" award. While still in his early twenties, Wolfram made a series of classic discoveries that helped lay the groundwork for the new sciences of complexity.

Not content with a traditional academic career, Wolfram created a software product called "Mathematica" in 1988, which rapidly became the standard system for technical computing used by millions of mathematicians, scientists, and engineers throughout the world. This achievement made Wolfram a multimillionaire and freed him to pursue a ten-year quest to develop a fundamentally new approach to the way that science models the natural world.

The essence of Wolfram's paradigm shift is that natural processes are best understood as an iterative series of interactive computations or algorithms rather than as the incarnations of abstract mathematical equations. This perspective offers startling new insights into a remarkable array of fundamental questions, including how biology produces complexity, how randomness arises in physics, and what space and time fundamentally are. In particular, the Wolfram approach may allow us to get a preliminary fix on a question of central importance to the Selfish Biocosm hypothesis: How complicated must the underlying cosmic program be in order to generate all the complexity we see around us, including life and intelligence?

Wolfram suggests that "beneath all the complex phenomena we see in physics there lies some simple program which, if run for long enough, would reproduce our universe in every detail."[6] This ultimate program, which Wolfram believes might be exceedingly brief, would function as the genetic code of the cosmos—a kind of "Software of Everything":

> [W]ith such a program one would finally have a model of nature that was not in any sense an approximation or idealization. Instead, it would be a complete and precise representation of the actual operation of the universe—but all reduced to readily stated rules. In a sense, the existence of such a program would be the ultimate validation of the idea that human thought can comprehend the construction of the universe. But just knowing the underlying program does not mean that one can immediately deduce every aspect of how the universe will behave. . . . [T]here is often a great distance between underlying rules and overall behavior. And in fact, this is precisely why it is conceivable that a simple program could reproduce all the complexity we see in physics.[7]

Wolfram's conjecture suggests another possible falsifiable implication of the Selfish Biocosm hypothesis. In order for the hypothesis to remain viable, it may be necessary to show that, in Wolfram's phrase, "the universe just follows a single, simple, underlying rule."[8] This rule would constitute the basic "constitution" or genetic code of the universe and the font of the life-friendly laws and constants of physics that prevail within it. Cosmic replication would consist of transmission of this simple underlying rule or algorithm to a new baby universe, where the algorithm could begin directing the process of constructing a new cosmos that will eventually evolve to a state closely resembling its mature "mother" universe in every material respect. The hypothesized simplicity and shortness of the underlying algorithm or rule would seem to be consistent with both ease and accuracy of transmission of the algorithm from "mother" universe to baby universe. A major future challenge will be to explore this assumption rigorously and to attempt to discern the irreducible minimum content of a rule or algorithm capable of generating a complex, life-friendly cosmos.

The Chemical Propensity for Life

A third area of fruitful future conjecture may center on efforts to generate falsifiable implications concerning the propensity of inanimate carbon-based polymers to self-assemble into living matter. The Selfish Biocosm hypothesis seems to imply that such a self-assembly propensity is preprogrammed into the physical laws that govern organic chemistry and, further, that such a tendency is relatively robust. The ongoing astrobiology research program sponsored by NASA may generate interesting evidence in the future that is relevant to this implication. (See sidebar, page 167.)

Plotting the Life Cycle of the Cosmos

A fourth area of future conjecture may involve an effort to determine whether the implications of the Selfish Biocosm hypothesis are consistent with a physically plausible "life cycle" for the universe. The hypothesis implies (1) that life and intelligence must be able to survive the physical challenges of all foreseeable cosmic eras and (2) that sufficient time remains in the projected life span of our cosmos to permit intelligent life to achieve a sufficient level of sophistication to serve as the duplicating machine that creates a new baby universe in the image of our own. Whether these two conditions can be satisfied, at least in theory, is presently unknown. Likewise unknown is the extent to which cultural factors (aversion or affinity of particular planetary populations for self-destructive thermonuclear conflict, for example) will affect the global persistence of life and intelligence in the cosmos. These are all areas of possibly fruitful future speculation and research.

Chapter Fifteen

A Scientifically Credible Strong Anthropic Principle

The received wisdom among traditional cosmologists is that the strong version of the anthropic principle—which provides that the laws and constants of inanimate nature encode a life-giving principle—is nothing more than a religiously inspired speculation. This book is one extended argument that it is possible to articulate a scientifically credible, fully naturalistic, and eminently testable version of the strong anthropic principle in the form of the Selfish Biocosm hypothesis. The key points of that argument shall now be recapitulated for the reader's convenience.

Hugh Everett and the Many Worlds Interpretation of Quantum Physics

The so-called Many Worlds Interpretation of quantum physics was developed by Hugh Everett and others to deal rationally with the odd philosophical implications of the Copenhagen Interpretation of quantum behavior (i.e., that physical reality is profoundly observer-dependent and that observer-participancy fundamentally structures that reality). Under the Many Worlds Interpretation, each act of quantum observation does not merely structure the microscopic reality being observed but actually splits the cosmos into an infinitely branching set of "many worlds," all of

(continued on next page)

Step 1

The starting point is an acknowledgment that the apparent fine-tuning of the laws and constants of inanimate nature that render them oddly friendly to carbon-based life cries out for a plausible naturalistic explanation—for a hypothesis with genuine explanatory power that is also falsifiable in the Popperian sense (i.e., capable of generating predictions that can be empirically tested). The challenge here is similar to that which faced Charles Darwin. Darwin sought a naturalistic paradigm that provided a nontheological explanation for the origin of species and for the appearance of design in nature. We seek a naturalistic paradigm that provides a nontheological explanation for the origin of the laws and constants of nature and that satisfactorily accounts for their peculiarly life-friendly qualities.

which continue to exist in various quantum states.

Step 2

The second step in the analysis is a recognition that the possibility of cosmological replication of one sort or another is a widely shared assumption among mainstream cosmologists and physicists, including Andrei Linde, Steven Weinberg, Alan Guth, Richard Gott III, Stephen Hawking, Neil Turok, Hugh Everett, Sir Martin Rees, and many others. Whether the proposed process of cosmic replication consists of eternal chaotic inflation (Linde and Weinberg), self-replication and self-creation of the cosmos by means of closed timelike curves (Gott), creation of baby universes in the lab (Guth and Rees), or quantum splitting of the universe into "many worlds" as a result of the act of measurement and observation (Everett), the key point is that there is a broad consensus that cosmic replication is a genuine physical possibility and thus that the creation of our particular universe through a Big Bang may not be a singular, one-shot affair. (See sidebar, page 175.)

Occam's Razor: The Virtues of Parsimony

Occam's Razor is scientific shorthand for the principle that the simplest and most parsimonious explanation for a natural phenomenon is to be preferred, all other things being equal. Named after its author, William of Ockham, a medieval philosopher and clergyman, the principle provides that "entities are not to be multiplied beyond necessity." This notion, which has been variously called the "law of economy" and the "law of parsimony," was invoked by Galileo to defend his hypothesis on the ground that it offered the simplest available explanation for the observed behavior of heavenly bodies.

Step 3

The third step in the analysis, which is closely related to the second, is to affirm that there is a broadly accepted consensus in the astrophysics community that cosmic replication may result in variation (either great or small) between the laws and constants of nature that prevail in a mother universe prior to the act of replication and those that prevail in a baby universe subsequent to replication. Some scientists suggest that the laws and constants may vary chaotically or randomly, while others (notably Lee Smolin) hypothesize that the laws and constants are heritable with only a small probability of variation from parent to offspring.

Step 4

The fourth step is a recognition, in consonance with the analysis of John von Neumann, that any self-replicating automaton (whether baby universe, human being, or minuscule bacterium) capable of reproducing itself and transmitting *heritable* traits to its offspring must, as a matter of inexorable logic, possess four key elements:

1. A *blueprint*, providing the plan for construction of offspring
2. A *factory*, to carry out the construction
3. A *controller*, to ensure that the factory follows the plan
4. A *duplicating machine*, to transmit a copy of the blueprint to the offspring

In the context of a cosmological replication hypothesis, one can assume that the physical laws and constants of our universe and its presumed progeny could conceivably constitute a von Neumann blueprint and that the universe at large could serve as a sort of von Neumann factory.

Step 5

The fifth step is to acknowledge that any serious assertion that the universe might be a self-replicating automaton capable of transmitting heritable traits to its progeny requires us to search diligently and with an open mind for candidate devices or natural processes that could conceivably play the roles of von Neumann's controller and von Neumann's duplicating machine in the context of a hypothesized process of cosmological replication and natural selection. These elements are needed to endow the hypothesized process of cosmological natural selection with the phenomenon of heredity (what Gerald Edelman calls "memory" in a broad sense).

Step 6

The sixth—and by far most controversial—step is to assert that highly evolved life and intelligence might conceivably be capable, in the far distant future, of serving as a cosmic von Neumann duplicating machine and that the natural processes of emergence hypothesized by evolutionary and complexity theory might serve as a von Neumann controller in this context. While far from certain, this assertion finds strong support in the theoretical speculations of a number of prominent scientists (including John Barrow, Frank Tipler, Alan Guth, Edward Harrison, Freeman Dyson, and others) that sufficiently evolved life and intelligence could conceivably exert a global effect on the overall state of the cosmos, much as life has reshaped the atmosphere of the Earth over the eons. At the very least, this possibility is widely recognized by most cosmologists to be within the realm of possibility. As Sir Martin Rees puts it, "We can't predict what role life will have carved out for itself ten billion (or more) years hence: it could be extinct; on the other hand, it could have evolved to a state where it can influence the entire cosmos, perhaps even invalidating [the] forecast [of indefinite cosmic expansion, leading eventually to eternal cosmic stagnation]."[1] Moreover, this assertion is, as we have seen, eminently falsifiable, thus qualifying as a genuinely scientific speculation.

Step 7

The seventh and final step in articulating the Selfish Biocosm hypothesis is to propose that the suggested scenario offers an encompassing, fully naturalistic, and reasonably parsimonious explanation for the oddly life-friendly laws and constants of nature that prevail in our cosmos. (See sidebar, page 176.) Under the theory, the laws and constants are life-friendly precisely *because* they were deliberately engineered by advanced intelligent life-forms in a prior cosmic cycle to endow our universe with the capability for life-mediated reproduction. The laws and constants are thus strictly analogous in function to that of DNA in an earthly organism. In the terminology used by pioneering quantum theorist Erwin Schrödinger in his pithy little book *What Is Life?* the laws and constants of inanimate nature are, under the tenets of the Selfish Biocosm hypothesis, both "code-script" and the instruments that "brin[g] about the development they foreshadow. They are law-code and executive power—or to use another simile, they are architect's plan and builder's craft—in one."[2]

The foregoing explication of the Selfish Biocosm hypothesis constitutes, I submit, a scientifically plausible, reasonably parsimonious, and eminently falsifiable version of the strong anthropic principle, which asserts that the laws and constants of inanimate nature that prevail in our universe specifically encode a robust capacity to generate life and increasingly competent intelligence. They do so, in an ontological sense, not because the universe exists to generate life and mind (or, in a parochial sense, to serve the interests of mankind), but for precisely the opposite reason: because life and mind arise in order to permit the universe to regenerate and reproduce itself. Life and intelligence serve the needs of the cosmos, not vice versa. Indeed, it is precisely because cosmic replication is the hypothesized utility function of the overall ontogenetic and replication process that the theory is called the "Selfish" Biocosm hypothesis.

This vision of the cosmos is the exact opposite of an anthropocentric conception. To paraphrase John Kennedy's famous admonition, the theory asks not what the universe can do to serve the interests of life and intelligence (including those of humanity), but rather what life and intelligence can do to serve the "selfish" interests of the cosmos. In this sense, the Selfish Biocosm hypothesis may be, from the human perspective, a cold and disconcerting concept. In this respect, it resembles Darwin's great theory.

Part Six

The Biocosm and Humanity

Part 6 of this book explores the possible implications of the Selfish Biocosm hypothesis for both evolutionary theory and for our self-image as a species. Because the new hypothesis deals with topics that are at the heart of traditional religious belief systems, part 6 also attempts to make a cursory assessment of the likely interaction between such beliefs and the new ideas expressed in the book.

BIOCOSM

Chapter Sixteen

Climbing Darwin's Ladder

Degeneracy and Evolutionary Progress

It is an uncontestable fact that "higher" organisms (like humans) are often relatively impoverished when compared to their microbial ancestors in terms of biochemical dexterity. Microbes can engage in feats of metabolic legerdemain that far exceed the capabilities of mammals and reptiles. Indeed, the story of metazoan evolution is, in large part, a tale of the loss of ancestral biochemical capacities.

Does this imply that evolution as a whole is a story about degeneration rather than progress? Certainly not.

What should not be forgotten is that while individual species may,

(continued on next page)

It is perhaps premature to speculate on the broad philosophical implications of the Selfish Biocosm hypothesis. But those implications are so novel, so interesting, and so contrary to currently prevalent mainstream philosophical thinking that the risk seems well worth taking.

The first and most obvious philosophical revision forced by the hypothesis is a radical rethinking of the basic nature of the process of biological evolution. It is virtually an article of faith among mainstream philosophers of science that the evolutionary process should be viewed as a kind of random walk (Stephen Jay Gould analogizes it to a drunkard's stumbling gait[1]) in a theoretical space that might be defined as the "library" of all conceivable versions of living species. Random variation and the ensuing natural selection of fitter organisms that occasionally emerge as the result of such variation is the heart of conventional interpretations of Darwin's theory. "Fitter" in this context does not mean that a particular organism is necessarily quicker or smarter or more stealthy than its competitors, but only that its characteristics more accurately match the specific physical challenges of its immediate environment (the fitness landscape), thus endowing the fitter organism with superior prospects for reproductive success. Indeed, a degenerate organism like a successfully reproducing parasite that has actually lost functionality may be unquestionably "fitter" under this interpretation than its ancient ancestors, which possessed a more complete endowment of metabolic functions.

The canonical view that the evolutionary process lacks directionality and fails to exhibit any pattern that might be characterized as progressive or embodying a principle of meliorism is captured perfectly by the following excerpt from an online essay by the inimitable Stephen Jay Gould:

> There is no progress in evolution. The fact of evolutionary change through time doesn't represent progress as we know it. Progress is not inevitable. Much of evolution is downward in terms of morphological complexity, rather than upward. We're not marching toward some greater thing. The actual history of life is awfully damn curious in the light of our usual expectation that there's some predictable drive toward a generally increasing complexity in time. It that's so, life certainly took its time about it: five-sixths of the history of life is the story of single-celled creatures only. I would like to propose that the modal complexity of life has never changed and it never will, that right from the beginning of life's history it has been what it is; and that our view of complexity is shaped by our warped decision to focus on only one small aspect of life's history; and that the small bit of the history of life that we can legitimately see as involved in progress arises for an odd structural reason and has nothing to do with any predictable drive toward it.[2]

Under the Selfish Biocosm hypothesis, the evolutionary process is analogous not to a random walk through a theoretical library containing all possible variations of species, but rather to the purposeful climbing of a ladder that leads to discrete and discernible plateaus populated with species of growing complexity and competence. (See sidebar, page 183.) In particular, the hypothesis is inconsistent with the conventional wisdom among evolutionary theorists that the probability of the emergence of human-level and higher intelligence through the natural process of biological evolution was exceptionally small and that if the pageant of biological history were somehow rerun, nothing resembling human intelligence would be likely to emerge. According to the Selfish Biocosm hypothesis, the appearance of human-level and vastly higher intelligence is a highly robust phenomenon, virtually foreordained by the basic laws of physics.

over the course of evolutionary time, forfeit certain capabilities with which their aboriginal predecessors were endowed, the community of organisms within which a particular species coevolves may provide substitute capacities. The humblest example is the human intestinal tract. As anyone who has ever swallowed doses of the most powerful antibiotics can attest, digestion cannot proceed properly without the assistance of the tiny bacteria that populate our gut. The overall progress of evolution can be properly measured only by reference to the aggregate capacities of an entire biosphere—a kind of superorganism composed of vast populations of coevolved constituent organisms.

David Bohm's Implicate Order

David Bohm was a radically original thinker who dared to ponder the deep metaphysical implications of quantum physics. Another twentieth-century physicist, Richard Feynman, who probably did more than any other scientist to refine quantum theory, once confessed that he thought he could safely say that no one on Earth really understood quantum mechanics. But no one has tried harder than Bohm to understand the theory's philosophical implications.

Bohm took seriously the most counterintuitive teachings of quantum physics—that reality was fundamentally nonlocal (i.e., that a measurement conducted billions of light-years away could reach back instantaneously over vast expanses of time and space to influence the behavior of a tiny electron), that reality is profoundly statistical (i.e., that one cannot meaningfully describe individual quantum processes), and that subatomic entities like photons can be simultaneously waves and particles.

Bohm was convinced that there must be a

(continued on next page)

The hypothesis, if taken seriously, may provoke serious questioning of the evolutionary paradigm popularized by Gould and a renaissance of currently disfavored interpretations of evolution as progress, including those advanced by contemporary evolutionary theorists Edward O. Wilson, Robert Wright, and Gregory Stock. The hypothesis may also revive the discredited thinking of the deceased visionary Teilhard de Chardin and the French philosopher of evolutionary science Henri Bergson, who became famous for expressing the view that life is driven to ever greater levels of sophistication by a mysterious *élan vital,* or vital force. Let us consider how the Selfish Biocosm hypothesis might relate to each of these viewpoints as well as to the recently rehabilitated concept of "symbiogenesis" (which roughly translates into "evolutionary progress through symbiosis"), to the exciting new field of research known as artificial life, and to a new vision of evolution-as-computation recently put forward by Stephen Wolfram. Finally we will contemplate parallels between the Selfish Biocosm hypothesis and the controversial Gaia theory.

Stephen Jay Gould's Canonical Vision of Evolutionary Contingency

No biological theorist since Darwin more thoroughly dominated public debate about evolution than did Stephen Jay Gould. Despite the unpopularity among professional colleagues of his signature contribution to evolutionary theory—the concept that the evolutionary process is properly characterized as exhibiting a pattern of "punctuated equilibrium" (i.e., long-term species stability, interrupted at rare intervals by episodes of rapid macroevolution)—Gould managed by dint of verbal dexterity and sheer intellectual will to dominate the public understanding of Darwin's theory for decades, at least in the United States.

While Gould's notion of punctuated equilibrium has drawn heated criticism from the academic community,[3] his more basic assumption—that the process of evolution is utterly contingent and bereft of any tendency that might be described as embodying an arrow of progress—is virtually unchallenged by most mainstream biologists.[4] Because this assumption is so fundamental, so widely and uncritically accepted within the scientific establishment, and so utterly inconsistent with the Selfish Biocosm hypothesis, it is worth analyzing in some detail.

The essence of Gould's assumption about the nondirectionality of evolution is captured in an introductory paragraph to his book *Full House.* In that introduction, Gould asserts that the historical record reveals "the unpredictability and contingency of any particular event in evolution" and emphasizes that "the origin of *Homo sapiens* must be viewed as such an unrepeatable particular, not an expected consequence."[5] Beyond that specific argument, Gould contends that the evolutionary record as a whole furnishes the basis for "the general argument for denying that progress defines the history of life or even exists as a general trend at all. Within such a view of life-as-a-whole, humans can occupy no preferred status as a pinnacle or culmination. Life has always been dominated by its bacterial mode."[6]

From this rather modest premise—that humanity is not the culmination of evolution—Gould draws the extravagant conclusion that the emergence of human-level intelligence was an extraordinarily unlikely contingency, not in the least foreshadowed by the deep forces of natural selection:

> If we are but a tiny twig on the floridly arborescent bush of life, and if our twig branched off just a geological moment ago, then perhaps we are not a predictable result of an inherently progressive process (the vaunted trend to progress in life's history); perhaps we are, whatever our glories and accomplishments, a momentary cosmic accident that would never arise again if the tree of life could be replanted from seed and regrown under similar conditions.[7]

Perhaps—and perhaps not. If Gould's hyperbolically phrased conclusions were intended to state a *scientific* hypothesis, he surely would have proposed falsifiable implications of that hypothesis just as I did in

deeper reality that underlies quantum phenomena and that the manifestations of quantum physics are epiphenomena of this fundamental ontology, which Bohm called the "implicate order."[20] Tragically, the bold theorist died before he was able to fully develop this new paradigm.

the essay discussed earlier in this book, which suggested specific tests of the proposition that the emergence of ever greater levels of intelligence is a robust evolutionary phenomenon. But upon careful analysis, it is clear that what Gould has propounded is not a hypothesis but an extended tirade, not a scientific theory but an ideological proclamation. Here is the inimitable Gould, in high-dudgeon mode, fulminating against the very idea of progress:

> Freud was right in identifying suppression of human arrogance as the common achievement of great scientific revolutions. Darwin's revolution—the acceptance of evolution with *all* major implications . . . —has never been completed. Darwin's revolution will be completed when we smash the pedestal of ignorance and own the plain implications of evolution for life's nonpredictable nondirectionality—and when we take Darwin's topology seriously, recognizing that *Homo sapiens*, to recite the revised litany one more time, is a tiny twig, born just yesterday on an enormously arborescent tree of life that would never produce the same set of branches if regrown from seed. We grasp at the straw of progress (a desiccated ideological twig) because we are still not ready for the Darwinian revolution. We crave progress as our best hope for retaining human arrogance in an evolutionary world. Only in these terms can I understand why such a poorly formulated and improbable argument maintains such a powerful hold over us today.[8]

What is truly interesting about the obvious level of Gould's passion is not the scientific view he states—for no falsifiable hypothesis is put forward—but rather the degree to which it reveals inadvertently how quickly issues like the inherent likelihood of evolutionary progress (or lack thereof) mutate into inflammatory ideological battle slogans. This curious phenomenon demonstrates the power of such ideas with regard to our self-conception as a species and our vision of humanity's place in this universe.

The Visions of the Contrarians

Not all evolutionary philosophers share Stephen Jay Gould's dismissive attitude toward the notion that evolution reveals a pattern of progressive change. As Gould admitted, even his distinguished Harvard colleague Edward O. Wilson is a contrarian on this issue. As Wilson has written (in a passage cited contemptuously by Gould),

> Many reversals have occurred along the way, but the overall average across the history of life has moved from the simple and few to the more complex and numerous. During the past billion years, animals as a whole evolved upward in body size, feeding and defensive techniques, brain and behavioral complexity, social organization, and precision of environmental control. . . . Progress, then, is a property of the evolution of life as a whole by almost any conceivable intuitive standard, including the acquisition of goals and intentions in the behavior of animals. It makes little sense to judge it irrelevant.[9]

Wilson's views—and those of other thoughtful philosophers of biology discussed below—represent an alternative philosophical interpretation of the phenomenon of evolution, currently out of favor, which would likely be revived by broad acceptance of the Selfish Biocosm hypothesis. That alternative interpretation was summed up crisply and eloquently by Christian de Duve in his book *Life Evolving:*

> A major objective of this book has been to expose the fallacy of [the] "gospel of contingency," which is being preached in the name of science. The alleged scientific premises of this doctrine, as I have tried to show, are incorrect. Not, as some would have it, because there is "something else" shaping the direction of evolution, but because the natural constraints within which chance operates are such that evolution in the direction of increasing complexity was virtually bound to take place, if given the opportunity. Chance does not exclude inevitability.[10]

Robert Wright's Nonzero Theory

The work of evolutionary philosopher Robert Wright centers on an attempt to discern whether there is a perceptible direction—even a decipherable meaning—to human history. This portentous question has, of course, been posed by philosophers and despots, prophets and tyrants, historians and fanatics over the millennia. For philosopher Immanuel Kant and an unlikely set of intellectual allies (including G. W. F. Hegel and Karl Marx, religious mystic Pierre Teilhard de Chardin, and French thinker Henri Bergson), the answer was unquestionably yes. Life and history, they believed, embody a logical pattern of inexorable progress—the unfolding of what Bergson famously called an *élan vital,* a vital force.

Others have disagreed vociferously. As we have noted previously, Nobel laureate Steven Weinberg has proclaimed with extravagant pessimism that "the more the universe seems comprehensible, the more it also seems pointless." Following World War II, historians Karl Popper and Isaiah Berlin took up arms against notions of historical laws and directionality, proclaiming them not only wrong but actually dangerous inasmuch as they frequently harbor violence-prone ideologies like fascism and Naziism. This remained the dominant view among eminent historians and cultural anthropologists throughout the latter half of the twentieth century.

Robert Wright entered the fray on the side of Kant and his defenders in a remarkable book published in 2000 entitled *Nonzero: The Logic of Human Destiny.* Wright articulates the hypothesis that there is indeed an inner logic and a directionality to history and even to biological evolution: "The more closely we examine the drift of biological evolution, and, especially, the drift of human history, the more there seems to be a point to it all. Because in neither case is 'drift' really the right word. Both of these processes have a direction, an arrow. At least, that is the thesis of this book."[11]

Wright derives his historical "arrow" and the title of his book from game theory, which distinguishes between "zero-sum" games (where one contestant's win is inevitably the other's loss) and "non-zero-sum" games (the proverbial "win-win" scenario). The evolution of life as well as the progression of human history toward ever higher levels of cultural complexity, Wright proclaims, constitute one gigantic non-zero-sum game:

> In short, both organic and human history involve the playing of ever-more-numerous, ever-larger, and ever-more-elaborate non-zero-sum games. It is the accumulation of these games—game upon game upon game—that constitutes the growth in biological and social complexity that people like Bergson and Teilhard de Chardin have talked about. . . . Non-zero-sumness . . . is something whose ongoing growth, and ongoing fulfillment, defines the arrow of the history of life, from the primordial soup to the World Wide Web.[12]

Wright's hypothesis of ongoing progressive change rests in large part on case studies of convergent cultural evolution toward societal complexity. These studies are powerfully persuasive and undercut the canonical view held by politically correct anthropologists and historians that no culture is more "primitive" than any other.

The Selfish Biocosm theory is deeply consilient with Wright's hypothesis about history's progressive arrow. Indeed, the theory suggests that such an arrow is embedded in the very fabric of reality in the form of a kind of genetic code of the cosmos.

Gregory Stock's Metaman Theory

The notion that human society can be usefully analogized to a giant living creature has a long and distinguished history. The seventeenth-century English political philosopher Thomas Hobbes analogized all of society to a single living being—Leviathan—whose essence could best be comprehended when refracted through the prism of metaphor. Other great thinkers throughout history have employed the same heuristic device to analyze and better understand (albeit in an unscientific manner) the underlying nature and behavior of human communities. As Gregory Stock puts it in his book *Metaman,*

> The metaphor that human civilization is an immense living being has been with us since the ancient Greeks, its form changing as scientific and philosophic vocabularies have shifted. In the twelfth century, John of Salisbury, inspired by Aristotle's writings, likened society to a creature in which each class played its God-given role: the king was the head, the church the soul, judges and governors the eyes and ears, soldiers the hands, and peasants the feet. In the late 1800s, Herbert Spencer's *Principles of Sociology* drew a more detailed analogy between biological organisms and society by describing a "social organism." In the early 1900s, Pierre Teilhard de Chardin, a biologist and Jesuit priest, linked evolutionary ideas to the concept of a global social organism and discussed the growing union of humankind in both biological and spiritual terms as an evolutionary transition toward a divine state. He described something he labeled the "noosphere," which was essentially an evolving collective consciousness. The great biologist Theodosius Dobzhansky extended Teilhard de Chardin's biological ideas in the 1960s, partially separating them from their theological framework. These thinkers and others were reaching for a way of understanding the extraordinary phenomenon of human society—its power, its integration, its dynamism.[13]

The Selfish Biocosm hypothesis appears to be consistent with Stock's hypothesis that "Metaman"—the hypothetical superorganism encompassing all of human civilization on this planet as well as the artifacts of that civilization (which Richard Dawkins might describe as the superorganism's extended phenotype)—is far more than a metaphor. The Selfish Biocosm hypothesis asserts that the inexorable march of life and intelligence up the slopes of Mount Improbable (Dawkins's phrase again) is virtually predestined by the basic laws of nature. Clearly, the history of biological complexification demonstrates that one critical pathway up the mountain of statistically improbable progress is through combination and cooperation of formerly discrete living entities (the incorporation of mitochondria into eukaryotic cells is the classic example).

If life and intelligence are to make the quantum leaps necessary to achieve the capacities required by the Selfish Biocosm hypothesis, it seems clear that ever closer functional integration of the human/machine amalgam we perceive as modern technological civilization is virtually inevitable. It seems equally likely that an emerging technology now perceived as profoundly threatening to the dignity of humanity—human germline therapy—will be a key instrumentality facilitating that integration.

Gene-Culture Coevolution: Darwinism on Steroids

A disquieting point of intersection between the Selfish Biocosm hypothesis and contemporary bioethics about which I have previously written involves the issue of the propriety of "human germline therapy"—a controversial technology that a task force of the National Association for the Advancement of Science advised in 2000 should not be pursued, at least at the present time.[14] Human germline therapy consists of the alteration of the human genetic sequence in reproductive cells to either avoid a predicted disability in progeny or enhance a particular trait associated with a certain gene. Therapeutic germline changes are "inheritable" because the altered genome is passed on during reproduction. This contrasts with "somatic gene therapy," an experimental medical technique that has already reached the stage of clinical trials (sometimes with disastrous results), which attempts to use viral vectors or other insertion methods to essentially retrofit the genome of a developed organism with a replacement for a missing or defective gene.

The topic of human germline therapy is likely to emerge in the not too distant future as a societal flash point that may make the abortion debate seem tame by comparison. According to newspaper reports in the fall of 2001, in a somewhat bizarre preview of probable developments to come, the famous physicist Stephen Hawking called upon governmental leaders to be open-minded with respect to the possibility of enhancing human mental and physical capabilities through genetic engineering. Hawking purportedly argued that this might be necessary in order to equip humans with capabilities sufficient to permit them to compete on equal terms with the rapidly evolving computers and robots endowed with ever more sophisticated artificial intelligence!

However eccentric Hawking's call to arms may seem, it does underscore a key fact: increasing biotechnological sophistication makes human germline modification a genuine possibility. Moreover, this technology has the potential of speeding up the process of human biological evolution enormously by transforming it into a technologically mediated process of gene-culture coevolution—similar to but vastly more powerful and rapid than the process of "unconscious selection" that figured so prominently in Darwin's theory. Should it materialize—and there is every reason to believe that it will in the not too distant future—a process of human-germline-therapy-enabled gene-culture coevolution would indeed constitute "Darwinism on steroids."

Any hypothesis (including the Selfish Biocosm hypothesis) that predicts exponential increases in the capacity of intelligence and human (or transhuman) capabilities in the future appears to be consistent with the possibility that this ethically troublesome technology will eventually be deployed in order to achieve this outcome. (There is a detailed discussion of this possibility and its ethical implications in my essay for *Complexity* entitled "Genes Beget Memes and Memes Beget Genes: Modeling a New Catalytic Closure," a copy of which is included in appendix 1.[15])

The Philosophy of Pierre Teilhard de Chardin

No scientific treatise in recent memory has encountered the kind of intense derision and contempt from the scientific establishment that was generated by the posthumous publication in 1955 of Pierre Teilhard de Chardin's *The Phenomenon of Man.*[16] Teilhard was a distinguished paleontologist and a Roman Catholic priest who, like many scientists, wished to cap his illustrious career with an expansive and frankly speculative reflection on the broad perspective he had gained from a lifetime of narrowly focused research. However, Teilhard confronted a barrier that many of his professional colleagues did not face: he was a priest, subject to oversight and supervision by the Roman Catholic Church.

The Phenomenon of Man was judged sufficiently heretical by the Catholic hierarchy that its publication was not permitted during the author's lifetime—a factor that undoubtedly contributed to the book's subsequent notoriety and to the author's heroic image as a victim, like Galileo, of Catholic Church censorship.

Following publication, *The Phenomenon of Man* (which Teilhard insisted on labeling as genuine science rather than theology or metaphysics) provoked a raft of hostile comments from the scientific establishment. The reaction of Peter Medawar, a distinguished British biologist, was typical. Medawar excoriated the book as "nonsense, tricked out by a variety of tedious metaphysical conceits" and allowed as how "it cannot be read without a feeling of suffocation, a gasping and flailing around for sense." The only saving grace of *The Phenomenon of Man*, according to Medawar, was that the "author can be excused of dishonesty only on the grounds that before deceiving others he has taken great pains to deceive himself."[17] The French Nobelist Jacques Monod confessed that he was "struck by the intellectual spinelessness of (Teilhard's) philosophy."[18]

While dismissed by most biologists (with the notable exception of biochemist Harold J. Morowitz, who previously served as editor of the Sante Fe Institute's *Complexity* journal and who proudly proclaims Teilhard to be his "role model" in "speculative scholarship"[19]), Teilhard's strange vision of the future has achieved resonance both with the public and with an unlikely scientific cohort: physicists and cosmologists investigating the various versions of the anthropic cosmological principle. Why is this?

There is something eerily familiar about the way Teilhard describes the force that propels biological complexification, which he calls by the rather mystical name of "radial" energy. Radial energy, which is to be contrasted with the more familiar "tangential" energy (the type of energy measured by a physicist's instruments), is a deep ordering force—a kind of primal antichaos—that drives the entire biosphere to higher and higher levels of complexity and sophistication. Radial energy constitutes an integrative force equal if not superior to the entropic force codified in the Second Law of Thermodynamics. Unlike tangential energy, whose dissipation increases inexorably with the passage of time, radial energy will, according to Teilhard, become steadily more concentrated, pushing forward the evolution of man and whatever transhuman living creatures may evolve in the future.

There are clear points of conceptual convergence between Teilhard's vision and the thinking of complexologists like Stuart Kauffman as well as quantum physicists like David Bohm, who postulated the existence of an "implicate order" that underlies and gives structure to both quantum and classical reality. (See sidebar, page 184.) What Teilhard calls radial energy, Kauffman would call the universe's innate propensity for self-organization, its irrepressible instinct for order. Where Teilhard parts company with the freethinking Kauffman and other theoretical biologists is in articulating a mystical vision of the Omega Point, discussed in chapter 1 and described as follows by John Barrow and Frank Tipler in *The Anthropic Cosmological Principle*:

> What is the goal of mankind, according to Teilhard? Just as non-sapient life covered the Earth to form the *biosphere*, so mankind—thinking life—has covered the Earth to form what Teilhard terms the *noosphere*, or cogitative layer. At present the noosphere is only roughly organized, but its coherence will grow as human science and civilization develop, as "planetization"—Teilhard's word—proceeds. Finally, in the far future, the radial energy will at last become totally dominant over, or rather independent of, tangential energy, and the noosphere will coalesce into a super-sapient being, the *Omega Point*. This is the ultimate goal of the tree of life and of its current "leading shoot," *Homo sapiens*.[21]

It is easy to see why this arresting vision of the future has captured the imagination of Internet visionaries. These techno-nerds see in the contemporary technological miracle of Internet connectivity, which weaves together both the biological and artificial neurons of the planet in an ever tightening embrace, the nebulous outline of an emerging transhuman nervous system, a global brain, an Omega Point in the throes of birth.

For our purposes, the important point about Teilhard's analysis is that it is forthrightly and unashamedly teleological in emphasis. For Teilhard, attainment of the Omega Point is what evolution is all about. And progress toward the Omega Point is just that—*progress.* While life must grope its way through trial and error to ever higher plateaus of complexity, there is no doubt left by Teilhard's approach that evolution exhibits an inherent upward directionality. This aspect of his philosophy fits quite nicely with the Selfish Biocosm hypothesis.

Symbiogenesis: Evolution Through Cooperation

A breathtakingly original reconception of the evolutionary process is now more than half a century old—and still not quite off the ground. This novel approach—called "symbiogenesis"—could be given a new lease on life by the Selfish Biocosm hypothesis. The essence of the symbiogenesis model is that the process of macroevolution occurs, at least in large part, when formerly independent organisms begin to cooperate in living communes where they can pool their various talents for mutual advantage. This theory, despite its obvious conceptual appeal and factual foundation in the world of nature, has always been controversial for reasons that are not entirely clear.

The classic instance of symbiogenesis is the appearance of metazoans—multicelled animals made up of a multitude of single living cells. The population of single-celled organisms that are grouped together to form a multicellular organism obviously cooperate and divide up the labor of daily living—a point that seems almost too obvious to debate. The evidence that symbiogenesis has profoundly shaped evolutionary history is, in fact, pervasive at all levels of emergence—from the appearance of eukaryotic cells (through the incorporation of formerly independent mitochondria and chloroplasts as cellular organelles) to the clear indication that the emergence of advanced mental functions is an artifact of human social interaction (or gene-culture coevolution, as Edward O. Wilson puts it).[22]

An extremely important aspect of the symbiogenesis approach is that it may hold the key to understanding the dynamics that will propel the next great evolutionary leap forward, which is likely to be fueled by a technology-mediated (and thus vastly accelerated) version of Wilson's hypothesized gene/culture coevolution (see my *Complexity* essay reprinted in appendix 1 for a detailed analysis).[23] This is because such a coevolutionary process may be best understood as a dynamic fusion of the operation of disparate sets of replicators—genes and memes—furthering overlapping replication objectives or utility functions.

This process could turn out to be a hyperspeed version of the coevolutionary process that yielded human-level intelligence. The key theoretical challenge, described as follows by Wilson and Harvard professor Charles J. Lumsden, is to construct a rigorous theory accounting for that baseline coevolutionary process:

> The proper account of mind and culture must depend on a theory of gene-culture coevolution, which is the understanding in modern scientific terms of the great circuit of causation that runs from the genes to brain architecture and the epigenetic rules of mental development, then to the formation of culture, and finally back to the evolution of the genes through the operation of natural selection and other agents of evolution. We have offered evidence suggesting that hereditary and environmental influences cannot be cleanly separated. They work together to create the whole brain, mind, and culture. At the next higher level, genetic and cultural evolution are also permanently linked, a circumstance that led to our choice of the expression *coevolution.* We suggest that the human species is the product of a unique episode of gene-culture coevolution that has accelerated the growth of the brain and mental capacity during the past two million years. So potent is the process that some of the genetic advance in symbolic reasoning and language could have occurred during the past 50,000 years or less. It might be continuing right into historical times.[24]

If the premises of sociobiology, as articulated above by Wilson and Lumsden, are even close to correct, the future coevolution of genes and memes may turn out to be best understood as an instance of nascent symbiogenesis—that is, a dynamic process by which two or more sets of disparate replicators maximizing disparate replication objectives are able to join together to synergistically maximize their respective survival skills.

If the basic concept of symbiogenesis is so undeniably attractive, why is it still regarded as faintly disreputable by mainstream Darwinists? It is a difficult question to answer, other than to say simply that there are styles of thought that are popular in mainstream science in any particular historical era. Symbiogenesis may be simply out of style. However, the ascendance of the Selfish Biocosm hypothesis might herald a new dawn for this neglected paradigm. Why? Because that hypothesis is inescapably a scenario about evolution through cooperation and synergy, of the emergence of supermind through the fusion of the power of many individual minds.

Artificial Life: Simulating Evolution and Emergence

The great scientific weakness of Darwin's theory of evolution is that the full multi-billion-year evolutionary process—leading from prebiotic organic polymers to the emergence of human-level intelligence —happened, so far as we know, only once and only on planet Earth. There is no second set of data points and thus no way to rerun nature's most magnificent experiment. Or is there?

The promise of an exciting new field of computer research known as "artificial life" (also called "a-life") is that it may offer the opportunity to duplicate *in silica* (that is to say, in the memory and logic circuits of a computer or computer network) the evolutionary process—to, in effect, rerun the tape of life's history and thus to determine experimentally whether a generic version of this process robustly yields ever greater levels of complexity or whether the emergence of complexity is an exceptionally rare occurrence, as Stephen Jay Gould contended.

The experimental subjects in artificial life research are not actual biological creatures but bits of computer software (known as software "agents") that are programmed with relatively simple sets of behavioral traits. These traits might include nutrient-seeking behavior, patterns of hostility to or cooperation with neighboring artificial creatures, and the capacity to "learn" with the aid of implanted genetic algorithms from encounters with their artificial environments or competitors.

Once "created," these artificial creatures are set loose in virtual fitness environments that are also programmed with a set of relatively simple rules and that exist solely in the memory space of a computer or a network of computers. The artificial creatures then compete or cooperate on this virtual fitness landscape to further the simple objectives with which they have been programmed. The fundamental objective of many artificial life experiments is to determine whether the ensemble of artificial creatures that populate a virtual fitness environment exhibit "emergent" behaviors (i.e., complex patterns that could not have been predicted in advance on the basis of the simple preprogrammed rules that constitute their "genetic" endowment) and, if so, what the ensuing patterns of emergence reveal about the innate tendency of this highly abstracted version of the evolutionary process to generate complexity.

Artificial life as a scientific discipline is still in its infancy, but it is already sufficiently promising to have attracted serious attention from NASA astrobiologists who are considering incorporating it within the portfolio of scientific specialties needed to deal with the theoretical challenges raised by the possibility of "exobiology"—the possible existence of biological systems in extraterrestrial environments like Mars and Europa. The early results from artificial life experiments are quite interesting with respect to the issue of the evolutionary robustness of the emergence of complexity. For instance, a study published in 2000 by Japanese researchers Chikara Furusawa and Kunihiko Kaneko from the University of Tokyo concluded as follows:

> To sum up, our study has provided evidence that an ensemble of cells with a variety of dynamics and stable states (cell types) has a larger growth speed than an ensemble of simple cells with a homogeneous

> pattern, because of its greater capability [to] transport and share nutritive chemicals. Since no elaborate mechanism is required for the appearance of this complex cell system, our results suggest that complexity of multicellular organisms is a necessary course in evolution, once a multicellular unit emerges from cell aggregates.[25]

The great promise of artificial life research is not that it will one day generate artificial organisms that are truly alive—though a number of a-life scientists are beginning to believe that this is a distinct possibility—but that it will allow us to use sophisticated computer programs to study, probe, and simulate the internal dynamics of the evolutionary process as never before. This will hopefully broaden our understanding of not only *whether* the emergence of complexity is a robust evolutionary phenomenon but also the microscopic causal issues of *how* and *why* complex biological phenomena emerge from the interaction of simpler elements. If this promise is fulfilled, then the study of evolution will be utterly transformed into a truly experimental science like physics or chemistry, and the role of abstract theoreticians like Gould (who prided himself on the fact that he was not particularly quantitative and not a great experimentalist) will be irrevocably diminished.

Stephen Wolfram: Evolution as Computation

We have already encountered the mathematician and physicist Stephen Wolfram in the context of our analysis of further possible falsifiable implications of the Selfish Biocosm hypothesis (i.e., the issue of the minimum informational content of the "genetic code" for a baby universe). Wolfram has also put forward some interesting speculations about the nature of biological evolution that relate to the implications of the Selfish Biocosm hypothesis with respect to that process.

In the introduction to his book *A New Kind of Science*, Wolfram has this to say about evolution:

> The Darwinian theory of evolution by natural selection is often assumed to explain the complexity we see in biological systems—and in fact in recent years the theory has also increasingly been applied outside of biology. But it has never been at all clear just why this theory should imply that complexity is generated. And indeed I will argue . . . that in many respects it tends to oppose complexity. But the discoveries [I have made] suggest a new and quite different mechanism that I believe is in fact responsible for most of the examples of great complexity that we see in biology.[26]

And just what is that mechanism? It is, in Wolfram's view, a subset of the mechanism capable of generating all the complexity and diversity of the universe—a simple iterative program, in this instance encoded in the superlatively efficient source code we call the genetic sequence, which effortlessly generates biological complexity, essentially because complexity is easy to generate:

> On the basis of traditional biological thinking one would tend to assume that whatever complexity one saw must in the end be carefully crafted to satisfy some elaborate set of constraints. But what I believe instead is that the vast majority of the complexity we see in biological systems actually has its origin in the purely abstract fact that among randomly chosen programs many give rise to complex behavior.[27]

The role of natural selection is not to *generate* the creative complexity that propels the onward march of the evolutionary process but rather to *prune* its exuberant expression back to a state of maximum simplicity. Says Wolfram, "So what is the role of natural selection in all of this? My guess is that as in other situations, its main systematic contribution is to make things simpler, and that insofar as things do end up getting more complicated, this is almost always the result of essentially random sampling of underlying programs—without any systematic effect of natural selection."[28]

What does this mean in the context of the Selfish Biocosm hypothesis? Perhaps this: Biological evolution may be an integral part of the universal phenomenon of reality-as-computation. The basic logic that may be capable of regenerating the entire complexity of the cosmos from a simple iterative program—the so-called Software of Everything—may be embedded in animate as well as inanimate nature.

What this might mean is that there are underlying universal laws of complexity, still largely undiscovered, that are locked into the very logic of the universe and that endow cosmic processes and their constituent subroutines with an inherent tendency to produce cascading phenomena of increasing complexity. These phenomena might include the appearance and evolution of life and the emergence of ever greater intelligence. Physicist Paul Davies explains it thusly:

> The hope of many complexity theorists is that some sort of self-organizing physical processes could raise a physical system above a certain threshold of complexity at which point these new-style "complexity laws" would start to manifest themselves, bestowing upon the system an unexpected effectiveness to self-organize and self-complexify. The result would be a series of transitions that ratchet the system abruptly up the complexity ladder. Under the bidding of such laws, the system might be rapidly directed towards life. If that is correct, it would mean that life is not so much written into the laws of physics as built into the logic of the universe.[29]

Gaia Theory and the Selfish Biocosm Hypothesis

With the exception of Pierre Teilhard de Chardin's *The Phenomenon of Man*, no recent theoretical contribution to biology has generated greater controversy than the "Gaia" theory of the terrestrial ecosystem proposed by NASA atmospheric chemist James Lovelock[30] and presented in refined form by Lovelock and evolutionary theorist Lynn Margulis.[31]

The story of the origin of the theory is quite interesting. Lovelock was working on an analysis of the Martian atmosphere in an attempt to ascertain the likelihood of life on Mars. He discovered that the atmosphere of Mars was very close to chemical equilibrium, in contrast to the atmosphere of Earth, which is very far from chemical equilibrium. It was the presence of life, Lovelock concluded, that explained this key difference between the atmospheres of the two planets.

His conclusion that the biosphere of a planet could effect large-scale changes in the chemical composition of an entire planetary atmosphere was initially highly controversial but is now broadly accepted. Indeed, NASA has concluded that far-from-equilibrium planetary atmospheres are key markers of living planets—and perhaps the only biomarker that will be evident from light refracted through the atmospheres of so-called exoplanets circling distant stars.

It was the further refinement to this theory that generated the strongest controversy. As Norwegian scientist Keith Downing put it in an essay summarizing the history of the Gaia theory, "Lovelock and Lynn Margulis felt that the biota had an even stronger role: they could not only affect the planet, but could do so in a manner that was beneficial to life. In short, the biota indirectly regulate planetary conditions within a window of survival that is largely defined by their own physiologies."[32]

This refinement invited the interpretation of the entire set of earthly biota as a vast homeostatic system—essentially a single planetary superorganism (Gaia). This somewhat daring interpretation provoked a violent counterreaction from traditional Darwinists, who derided Lovelock and Margulis as softheaded sentimentalists and unscientific panderers to popular cravings for a holistic explanation of the unity of life on the planet. Nonetheless, as Downing notes, key elements of the theory were virtually impossible to refute:

> [The] radical [version of the Gaia theory] has evoked cries of "teleology" and "pop ecology" from a host of renowned scientists, but putting aside ancient prejudices and stale dogma, one sees a variety of inter-

> esting examples where the biota appear to play a major role in making the planet more livable. These include the regulation of local climate by algae, the control of global temperatures by photosynthetic organisms, the maintenance of relatively constant marine salinity and nitrogen-phosphorous ratios by aquatic biota, and the emergence of efficient recycling loops among diverse microbial species.[33]

The principal barrier to acceptance of the broadly stated Gaia hypothesis was its seeming inconsistency with the basic Darwinian mechanism. How could a fitness environment that emerged from a relentless struggle for survival and that rewarded the fittest organisms with the ultimate prize of reproductive success somehow metamorphose into a metasystem that was essentially benign and favorable to the continued existence of all competitors? It was, as Downing reveals, a daunting intellectual conundrum:

> As evidence of these [Gaian] phenomena accumulate, many natural scientists have accepted Gaia's essence: life begets life. However, many neo-Darwinians remain skeptical, since the evolutionary origins of ecosystem-level homeostatic loops with biotic components are difficult to envision, and are in fact counter to the competitive, survival-of-the-fittest views of natural selection. Gaian interactions involve the coordination of many biological, chemical and physical activities, as illustrated by the networks of diverse microbes involved in the global chemical cycles. Reconciling the emergence of coordinated, multi-species, distributed controllers with neo-Darwinian evolution is no simple task, even when coordinated strategies are clearly the best for all organisms.[34]

A serious effort is now under way, using the latest software tools of artificial life research, to attempt to model the emergence of a global homeostatic system—the metabolism of Gaia, if you will—from the interactions of constituent biotic subsystems. If it succeeds, this effort will not only triumphantly vindicate the Gaia hypothesis but provide strong support for a key premise of the Selfish Biocosm hypothesis: that the natural forces of evolution, emergence, and self-organization inexorably propel biotic systems, wherever they may exist, to ever higher levels of complexity and integration. It is interesting that two theories that developed from very different starting points—the Gaia hypothesis arose from an effort to explain the far-from-equilibrium status of the Earth's atmosphere, while the Selfish Biocosm hypothesis is an attempt to account for the oddly life-friendly laws of physics that prevail in our cosmos—should converge on this key point.

Darwin's Ladder

As Paul Davies noted in his book *The Sixth Miracle*, many scientists and nonscientists are drawn almost unconsciously to the metaphor of evolution as a ladder or, better yet, a high-powered escalator.[35] The real source of opposition to this metaphor is ideological, not scientific. It is because traditional Darwinists like Stephen Jay Gould believe that the idea of progress in evolution is a surrogate for religious beliefs that they oppose the concept so mightily. Gould exalted the notion of life's pointlessness as a stringent antidote to what he views as the dangerously seductive view that evolution "suppl[ies] the most desired ingredient of Western comfort: a clear signal of progress measured as some form of steadily increasing complexity for life as a whole through time."[36] In his view the appearance of advanced life-forms capable of comprehending the physical laws that guide cosmic processes was not foreordained but is instead the result of sheer chance: "We are glorious accidents of an unpredictable process with no drive to complexity, not the expected results of evolutionary principles that yearn to produce a creature capable of understanding the mode of its own construction."[37]

The Selfish Biocosm hypothesis yields precisely the opposite conclusion, not because that hypothesis harkens back to a nostalgia for pre-Darwinian notions of supernatural design, but because the heart of the hypothesis is that the progressive, self-complexifying processes of biological evolution and emergence, yielding ever more complex life and ever more competent intelligence as a natural, predictable and robust consequence, are written into the very logic of the universe and its physical laws and constants.

This view of life may have been anathema to Gould but it is strikingly reminiscent of that expressed in the concluding paragraph of Darwin's *Origin of Species*:

> It is interesting to contemplate a tangled bank, clothed with many plants of many kinds, with birds singing on the bushes, with various insects flitting about, and with worms crawling through the damp earth, and to reflect that these elaborately constructed forms, so different from each other, and dependent upon each other in so complex a manner, have all been produced by laws acting around us. These laws, taken in the largest sense, being Growth with Reproduction; Inheritance which is almost implied by reproduction; Variability from the indirect and direct action of the conditions of life, and from use and disuse: a Ratio of Increase so high as to lead to a Struggle for Life, and as a consequence to Natural Selection, entailing Divergence of Character and Extinction of less-improved forms. Thus, from the war of nature, from famine and death, the most exalted object which we are capable of conceiving, namely, the production of the higher animals, directly follows. There is grandeur in this view of life, with its several powers, having been originally breathed by the Creator into a few forms or into one; and that, whilst this planet has gone cycling on according to the fixed law of gravity, from so simple a beginning endless forms most beautiful and most wonderful have been, and are being evolved.[38]

Chapter Seventeen

Mind and Supermind

Genetic Algorithms and Evolutionary Computation

What if the design and fabrication of sophisticated software were no longer subject to the limitations of the human intellect? What if highly advanced software artifacts could somehow be coaxed into an evolutionary cycle that would employ the Darwinian tool of natural selection to generate ever more capable software specimens on an astonishingly compressed timescale? And what if such software could "learn" from prior evolutionary mistakes and thus benefit from a kind of Lamarckian adaptation that is absent from the natural world of biological evolution?

(continued on next page)

This chapter and the next venture into the realm of metaphysics—treacherous territory for any theorist who would desire to have his or her scientific speculations taken seriously. This chapter will examine the nature and possible enabling mechanisms of one of the developments predicted in my Rio de Janeiro paper: the emergence of transhuman intelligence. This emergence, it will be remembered, was one of the key falsifiable implications of the Selfish Biocosm hypothesis. It is, I believe, an implication whose validation is rather more imminent than certain other implications of the hypothesis.

The topic of the emergence and role played by mind in the cosmos occupies a vast literature, spanning neurobiology, philosophy, and even quantum mechanics. What concerns us here is not the broad sweep of this expansive body of thought and speculation but rather the narrow question of what these various sets of insights might be able to tell us about the next great leap upward in the ongoing ascendance of mind. While it may be true, as Freeman Dyson has speculated, that we are no more capable of appreciating the thoughts and emotions of the highly advanced life-forms that may inhabit the universe billions of years hence than a butterfly is capable of appreciating our human thoughts and feelings, it seems plausible that we should at least be capable of anticipating the thoughts that will occupy the minds of beings who will inhabit the next few rungs up on the evolutionary ladder that mind is steadily ascending. What follows are some tentative speculations about this daunting topic.

Supermind Will Build on What Has Gone Before

Just such a vision informs evolutionary computation and genetic algorithm research projects under way at a number of academic and defense institutions in the United States and abroad. A useful overview of this research is provided by a 2003 Scientific American *essay, "Evolving Inventions."*[2]

Perhaps the safest bet we can make is that whatever superintelligence succeeds human-level intelligence as the apogee of intellectual sophistication on Earth will build on the existing heritage of both biological and cultural evolution. This has been the pattern of evolution historically, and there is no reason to believe that it will not be repeated in the future. The foundation from which supermind is likely to evolve as the Selfish Biocosm scenario unfolds will doubtless include as cornerstones the obvious artifacts of advanced technological civilization—the Internet, the growing power and sophistication of computers—as well as the probable hybridization of biology and engineering, foretold in numerous works of science fiction about the creation of human/machine chimera known as cyborgs, which is beginning to emerge as a central theme of twenty-first-century technology.

The specific selectional pressures that will drive evolutionary progress are likewise present, at least in embryonic form. They include the vast range of exigencies that can be better managed through enhanced computational capabilities such as the following:

- Defense-related simulation of phenomena ranging from nuclear weapons explosions to the emergence of terrorist threats and the organization and dynamics of terrorist networks

- Computationally demanding tasks like modeling of genome expression, genome/proteome interaction, protein folding, simulation of the behavior of superstrings under various sets of assumptions, and the autonomous operation of unmanned spacecraft beyond the effective reach of remote control from Earth

- Such socially beneficial tasks as better weather forecasting, enhanced transportation planning, more accurate financial and market simulation, seamless switching and conversion from one kind of digital data to any other, development of the technology needed to produce cheap and abundant energy, and improved modeling of the biosphere in order to enhance environmental protection

These challenges present a pervasive set of evolutionary pressures, typically manifested as economic incentives, which are constantly pressing our society to improve its aggregate computational capacity.

Finally, the computational characteristics that will distinguish supermind from its human predecessor can most plausibly be described in terms of enhanced versions of existing human mental capabilities, which might well include

- enhanced processing speed;
- enhanced memory, both in terms of sheer quantity of content and accuracy;
- enhanced possibilities for neuronal networking.

None of these predicted enhancements seems particularly outlandish—or even controversial—if considered in the context of evolving machine intelligence.

The Emergence of Supermind: Self-Organization and Artificial Evolution

As George Dyson notes in *Darwin Among the Machines*, the concept of self-organization as a basic paradigm with which to explicate the mysteries of biological and technological evolution became exceedingly popular in the mid-twentieth century:

> Theories of self-organization became fashionable in the 1950s, generating the same excitement (and disappointments) that the "new" science of complexity has generated in recent years. Self-organization appeared to hold the key to natural phenomena such as morphogenesis, epigenesis, and evolution, inviting the deliberate creation of systems that grow and learn. Unifying principles were discovered among organizations ranging from a single cell to the human nervous system to a planetary ecology, with implications for everything in between.[1]

Whatever disappointments the paradigm might have generated, its legacy remains strong and viable in the fields of computation and communication. The innovations that resulted from the mid-twentieth century focus on self-organization as a conceptual model include modern air-defense computer control systems, most computer data networks in existence today, and what is perhaps the signature communications innovation of the late twentieth century, the Internet. All of these innovations grew, in one form or another, from an encompassing vision of nature as a self-organizing system.

As computer capacities grow exponentially more powerful in accordance with Moore's Law (the notion promulgated by the cofounder of Intel that the capacity of computation per unit of cost roughly doubles every eighteen months), the potential for self-organization within computational environments increases apace. Indeed, the sheer complexity of such environments (and of the software artifacts needed to navigate them successfully) implies that we will soon reach the limits of human capabilities to understand and intellectually dominate the remarkable self-constructing informational environment that some call the "infosphere," others label as the "telecosm," and a few crusty veterans still call "cyberspace." In the view of many software visionaries, when that limit is reached we will have only one choice if we wish to continue the process of software and hardware evolution: to empower computer software and hardware to, in effect, take charge of its own evolution. When this occurs, computer evolution will begin to recapitulate, on a vastly accelerated schedule, the process of biological evolution that yielded human-level intelligence.

Here is the scary news: the process has already begun. The task of coaxing software to emulate biological evolution is well under way at a number of leading research institutions around the world, including Los Alamos National Laboratory and the Santa Fe Institute in New Mexico. Research initiatives in such fields as artificial life, genetic algorithm creation, and evolutionary computation are already yielding exciting results and fueling the creation of an entire community of researchers intent upon the task of introducing Darwinian selectional processes into the world of software development. (See sidebar, page 207.)

The Soul of Supermind

In a stunning preview of what the future may hold, the physician and biologist Lewis Thomas predicted in an essay published in the *New England Journal of Medicine* in 1973 that if current trends of cultural and technological evolution persist, humanity and its artifacts will eventually coevolve into a single global intelligence. He stated, "All 3 billion of us are being connected by telephones, radios, television sets, airplanes, satellites, harangues on public-address systems, newspapers, magazines, leaflets dropped from great heights, words got in edgewise. We are becoming a grid, a circuitry around the earth. If we keep at it, we will become a computer to end all computers, capable of fusing all the thoughts of the world."[3]

Thomas's speculation echoes a poetic reflection, published more than 150 years ago, by the great American author Nathaniel Hawthorne on the occasion of the deployment of the telegraph, an innovation at least as momentous in its day as the Internet is in today's world: "By means of electricity, the world of matter has become a great nerve, vibrating thousands of miles in a breathless point of time. . . . The round globe is a vast head, a brain, instinct with intelligence. . . . It is itself a thought, nothing but thought, and no longer the substance which we deemed it."[4]

The final question we can ask about the triumphal culmination of the evolution of intelligence on Earth is this: Will that intelligence—that supermind—be recognizable as anything remotely resembling its human antecedent? Some observers say no. Lewis Thomas, for example, warned that "if we were ever to put all our brains together in fact, to make a common mind the way ants do, it would be an unthinkable thought, way above our heads."[5]

I am inclined to the opposite point of view. Just as we continue to display (in Carl Sagan's felicitous phrase) the shadows of forgotten ancestors, so too the essence of the human spirit may linger on as a kind of inexpungeable legacy in the souls of whatever forms of intelligence succeed us on this planet and in the universe. If so, it would seem that our responsibilities to the cosmos extend far beyond the brief tenure granted our species as *conquistadors* of Earth. While, as Freeman Dyson has opined, "there are questions about the universe as it may be perceived in the future by minds whose thoughts and feelings are as inaccessible to us as our thoughts and feelings are inaccessible to earthworms,"[6] it does not follow that we bear no responsibility for the shape and content of those future thoughts and feelings.

It seems clear that the phenomenon of *consciously directed evolution*—which arguably includes as a subset the entirety of human cultural and technological evolution—implicitly imposes ethical responsibilities on humanity beyond those imposed on any other species in the biosphere. As the great humanist and evolutionary philosopher Julian Huxley concluded in an essay entitled "The New Divinity," "This earth is one of the rare spots in the cosmos where mind has flowered. Man is a product of nearly three billion years of evolution, in whose person the evolutionary process has at last become conscious of itself and its possibilities. Whether he likes it or not, he is responsible for the whole further evolution of our planet."[7] The Selfish Biocosm perspective suggests that the possibility of infusing a moral purpose into the process of consciously directed evolution may impose a *transgenerational* moral imperative—

an obligation to build a moral as well as a technological foundation for the benefit of the lives and minds yet to come.

The vast saga of biological evolution on Earth is one tiny chapter in an ageless tale of the struggle of the creative force of life against the disintegrative acid of entropy, of emergent order against encroaching chaos, and ultimately of the heroic power of mind against the brute intransigence of lifeless matter. Through the quality and character of our contribution to the progress of life in this epic struggle, we shape not only our own lives and those of our immediate progeny but the lives and minds of every generation of living creatures down to the end of time. We thereby help to shape the ultimate fate of the cosmos itself.

Chapter Eighteen

A New Measure of Humankind

Stephen Hawking concluded his monumentally successful *A Brief History of Time* on this inspirational note: "If we do discover a complete theory, it should in time be understandable in broad principle by everyone, not just a few scientists. Then we shall all, philosophers, scientists, and just ordinary people, be able to take part in the discussion of why it is that we and the universe exist. If we find the answer to that, it would be the ultimate triumph of human reason—for then we would truly know the mind of God."[1]

In a subsequent book that takes its title from this famous passage (*The Mind of God*), physicist Paul Davies pondered the implications of the apparent fine-tuning of the universe that renders it so strangely and fortuitously life-friendly. Davies asked whether this odd feature of our cosmos might offer a pathway to eventual discovery of a final theory of the cosmos:

> Through science, we human beings are able to grasp at least some of nature's secrets. We have cracked part of the cosmic code. Why this should be, just why *Homo sapiens* should carry the spark of rationality that provides the key to the universe, is a deep enigma. We, who are children of the universe—animated stardust—can nevertheless reflect on the nature of that same universe, even to the extent of glimpsing the rules on which it runs. How we have become linked into this cosmic dimension is a mystery. Yet the linkage cannot be denied.[2]

The Categorical Imperative: Kant's Golden Rule

The German philosopher Immanuel Kant sought to formulate a supreme principle of morality that he described as the "categorical imperative." The classic formulation of Kant's categorical imperative is this: Act only on that maxim by which you can at the same time will that it should become a universal law.

Kant argued that there were at least four cogent formulations of this general principle that helped explicate its application from different ethical perspectives:

- *The Formula of the Law of Nature: Act as if the maxim of your action were to become through your will a universal law*

(continued on next page)

The best that Davies could do to explain the mystery was to assert that human existence and human thought are not accidental oversights in the cosmic scheme of things:

> What does it mean? What is Man that we might be party to such privilege? I cannot believe that our existence in this universe is a mere quirk of fate, an accident of history, an incidental blip in the great cosmic drama. Our involvement is too intimate. The physical species *Homo sapiens* may count for nothing, but the existence of mind in some organism on some planet in the universe is surely a fact of fundamental significance. Through conscious beings the universe has generated self-awareness. This can be no trivial detail, no minor byproduct of mindless, purposeless forces. We are truly meant to be here.[3]

The Selfish Biocosm hypothesis carries this analysis a giant step further, asserting that life and intelligence are, in fact, integral to the process of cosmological ontogenesis and replication, not in any mystical sense, but in a fully naturalistic way. In advancing this assertion, the hypothesis constitutes a daring thrust into territory that is traditionally the exclusive province of religion and philosophy.

Put simply, the Selfish Biocosm perspective requires a drastic revision in traditional monotheistic concepts of a supernatural deity as the sole creator of the cosmos. The new paradigm implies that the cosmos actually creates and renews itself as an enormous self-organizing and self-renewing system, and, further, that each living creature, at each juncture in the cosmic life cycle, is responsible for a small but possibly indispensable contribution to the overall process of cosmic growth, evolution, and eventual renewal. While puny and seemingly insignificant, humans and other life-forms throughout the universe all participate in a vast emergent process of inconceivable power and scope. The hypothesis suggests as well that we are, in a very profound sense, ethically responsible not only for the future evolution of our own and other species but, at least in some part, for the evolutionary fate of the cosmos.

of nature.

- *The Formula of the End Itself: Act in such a way that you always treat humanity, whether in your own person or in the person of any other, never simply as a means, but always at the same time as an end.*
- *The Formula of Autonomy: So act that your will can regard itself at the same time as making universal law through its maxims.*
- *The Formula of the Kingdom of Ends: So act as if you were through your maxims a law-making member of the kingdom of ends.*

It is this aspect of the hypothesis that has the deepest philosophical implications. Specifically, the hypothesis may have the potential to generate a new kind of environmental philosophy, to alert us to our ethical obligations to future generations and potentially to yield quasi-religious metaphysical implications. These possibilities are explored in the following pages.

Neo-Kantian Environmentalism

The environmental philosophy incipient in the Selfish Biocosm hypothesis is, I suggest, a kind of environmental ethic derived from Immanuel Kant's classic "categorical imperative," an ethical principle commanding that every individual should act on the assumption that each of one's actions could be translated into a universally applicable ethical precept. (It is essentially a restatement of the Golden Rule: Do unto others as you would have others do unto you. See sidebar, page 215.)

The notion that every creature, great and small, plays some indefinable role in an awesome process by which life gains hegemony over inanimate nature implies that every living thing is linked with every other bit of living matter in a joint endeavor—a kind of cosmic "Mission Impossible"—of vast scope and indefinable duration. We soldier on together—bacteria, people, extraterrestrials (if they exist), and hyper-intelligent computers—pressing forward, against all odds and the implacable foe that is entropy, toward a distant future we can only faintly imagine. But it is together—in a spirit of cooperation tempered by conflict—that we journey hopefully toward our distant destination. If, like Sisyphus, we are occasionally pained by the weight of the stone we are pushing uphill and if our task strikes us, at least sporadically, as futile and absurd, we can at least take comfort in the astonishing fact that every creature that ever lived and ever will live shares our existential plight.

The path that life will take on this vast journey will be shaped in subtle ways not only by chance but also, in the case of intelligently directed processes that are designed by sentient beings like ourselves, by conscious choice. Thus, to at least some limited degree, we and our descendants are the "intelligent designers" who are consciously and unconsciously shaping the fate of the universe. With apologies to Pogo, we have met the creator of the cosmos and the cosmic creator is us.

The role of a system of ethics and morality is to expose and enlighten individuals about the consequences of their actions and to admonish them to assume responsibility for them. If we come to truly believe that we are embarked on a cosmic mission of the greatest possible urgency and that our colleagues in that mission are the other denizens of the biosphere, then there may be a conceptual basis for a new kind of environmental ethic, which could be phrased as a variation on Kant's classic statement of the categorical imperative: Do unto other life-forms as you would have other life-forms do unto you.

To implant such a neo-Kantian environmental ethic in our culture may not only be the morally right thing to do from an ethical point of view but also a pressing imperative from the perspective of sheer human self-interest. If the predictions of the Selfish Biocosm hypothesis regarding the imminent emergence of transhuman intelligence turn out to be accurate, then we must earnestly hope that such an intelligence will treat us kindly—more kindly, in fact, than we have treated other sentient creatures that share Earth's biosphere with us.

Ethical and Legal Obligations to Future Generations

During the early years of the twentieth century a supremely gifted European physicist explained the mystery of the constancy of the speed of light by suggesting that time itself was inconstant, slowing to a crawl as nature's ultimate speed limit was approached or as the force of gravity grew sufficiently strong. Albert Einstein's counterintuitive concept of temporal variability is a fruitful metaphor for probing novel questions of temporal jurisprudence, by which I mean the broad range of issues covered by the topic of justice between generations, that are thrown into sharp relief by the Selfish Biocosm hypothesis. His revolutionary conceptual framework teaches us that what may appear to the naive understanding to be a self-evident postulate—in Einstein's case, the supposedly absolute and immutable pace of time—can, in fact, constitute the crucial variable.

Applied metaphorically in the context of intergenerational justice, Einstein's concept suggests that the broadly shared assumption that a currently living generation should be absolutely privileged to exercise political sovereignty with respect to governmental decisions whose lingering effects will persist for many generations may conceal profoundly important issues of justice and equal protection. The concept implies as well that time itself may have become the crucial moral axis in terms of which the fundamental fairness of such decisions should be assessed.

The set of concepts that comprise the Selfish Biocosm hypothesis hints that the topic of justice between generations may become one of the most important emerging frontiers of ethics and jurisprudence in the decades ahead. In a world beset by threats of global warming and rampant destruction of endangered species, it is difficult to conceive of an ethical topic more urgently in need of thoughtful and expeditious exploration. The Selfish Biocosm hypothesis adds to the philosophical urgency of this inquiry by suggesting that all human generations—from first to last—are engaged in a cosmic endeavor of utmost importance that relies upon a fragile continuity of intellectual and technological progress across a multigenerational expanse of time.

The basic concept of justice between generations is not new. In one of the last works published in his lifetime, Kant sought to answer the question of whether the human race is continually improving.[4] Discounting the difficulty of "depicting those events whose *a priori* possibility suggests that they will in fact happen,"[5] he inferred from the "*passion* or *enthusiasm* with which men embrace the cause of goodness"[6] that "the human race has always been progressively improving and will continue to develop in the same way."[7]

This optimistic prediction of progressive social improvement had one mildly disturbing corollary. As Kant had observed in an earlier work,[8] social progress does not occur in a single lifetime but rather "will require a long, perhaps incalculable series of generations, each passing on its enlightenment to the next, before the germs implanted by nature in our species can be developed to that degree which corresponds to nature's original intention."[9] Accordingly, it was

> disconcerting . . . that the earlier generations seem to perform their laborious tasks only for the sake of the later ones, so as to prepare for them a further stage from which they can raise still higher the structure intended by nature; and secondly, that only the later generations will in fact have the good fortune to inhabit the building on which a whole series of their forefathers (admittedly, without any conscious intention) had worked without themselves being able to share in the happiness they were preparing.[10]

This temporal injustice, while perhaps puzzling in the abstract, has never received serious philosophical attention, undoubtedly because, as ethical theorist John Rawls has observed, "it is a natural fact that generations are spread out in time and actual exchanges between them take place only in one direction. We can do something for posterity but it can do nothing for us. This situation is unalterable, and so the question of justice does not arise. . . . There is no way for later generations to improve the situation of the least fortunate first generation."[11]

Although the unidirectionality of time precludes any inquiry into whether the advantages enjoyed by presently living generations are unjust in comparison to the privations suffered by their predecessors, it does not follow that all consideration of the subject of "justice between generations"[12] is bound to be fruitless. It is the irreversibility of the past that moots the question of whether those living in the present are behaving fairly toward their ancestors. No such barrier prevents inquiry into whether the present generations are obliged to behave fairly toward future generations.

Indeed, that topic has received serious, if limited, attention from political theorists. Rawls, for instance, approaches the question of justice between generations from the perspective of a contractarian theory of distributive justice. Rawls's contractarian theory is based on an explication of the moral implications of an imagined original social contract between hypothetical parties creating a new society who are unaware of where they will end up in the ensuing social order created by this contract. His theory is a recent example of thinking on this topic, which has also been dealt with by Karl Marx and Georg Wilhelm Friedrich Hegel as well as American revolutionary thinkers like Thomas Paine and Thomas Jefferson. Parties to a hypothetical social contract, Rawls speculates, would choose to allocate present resources so as to both "preserve the gains of culture and civilization, and maintain intact those just institutions that have been established [and] put aside in each period of time a suitable amount of real capital accumulation"[13] in order to further an eventual "full realization of just institutions and the fair value of liberty."[14] They would do so, Rawls theorizes, because parties in the original position are motivated not merely by self-interest but also by an altruistic desire to benefit their descendants for at least two generations.[15]

For Rawls, the concept of intergenerational fairness is not an incidental epiphenomenon of his contractarian theory of justice but tightly linked with its basic postulates. Rawls's notion of distributive justice is based in part upon the principle that each generation accumulates certain savings for the benefit of future generations (Rawls calls this the "savings principle"). The process of accumulation benefits each succeeding generation after the first. The seven principal features of intergenerational distributive justice identified by Rawls are as follows:

1. Savings principle: the process of accumulation benefits each succeeding generation after the first.

2. The entire course of accumulation aims at a just state of society.

3. No generation is subordinate to any other.

4. Justice does not require early generations to save excessively so that later ones are simply more wealthy. Savings is rather demanded as a condition of bringing about the full realization of just institutions and the fair value of liberty.

5. Since it is a fact that generations are spread out in time, it is not unjust that each generation can do something for posterity but posterity can do nothing for its predecessors. Similarly, it is not unjust that the very first generation must save and begin the accumulation process, but never receives any of the benefits of that savings. These are simply unalterable facts and, therefore, are not unjust.

6. The motivation to effectuate this savings principle is premised on the assumption that each generation should care at least for its immediate successors (at least two generations ahead).

7. The just savings principle is defined from the standpoint of the least advantaged group in each generation.[16]

In sum, a "just" policy toward future generations is one that would be chosen by sympathetically motivated parties to a purely hypothetical original social contract. Rawls's theory, like other jurisprudential attempts to deal with the issue of justice between generations, centers on an exploration of the logical ramifications of an intuitive sense of moral obligation. It is indifferent to the actual course of historical events and, accordingly, unconcerned with the possibility that an existing legal system might, either now or at some point in the future, incorporate a source of actual obligation toward future generations.

Yet such a possibility should not be discounted. If Rawls's theory accurately captures a pervasive sense of moral obligation to posterity, then it would seem to follow that there is at least a possibility that some societies or cultures, either now or in the future, would be inclined to translate principles of intergenerational fairness into positive law. Moreover, it would seem most likely that this could occur in a social order whose actual historical origin bore a resemblance to the hypothetical process of original consensus that is the foundation of contractarian theory *and* where the advent of science and technology permits at least an educated guess about the probable future impact of current technological patterns. The possibility certainly exists that the Selfish Biocosm hypothesis—which offers a perspective on the evolution of society that is inherently and irrevocably multigenerational—might help catalyze the transformation of abstract philosophical principles of intergenerational fairness articulated by ethicists like John Rawls into positive law.

The Cosmic Imperative of Altruism

The possibility of altruism would seem to be disallowed by the pitiless struggle for survival that is the basic postulate of Darwinian theory. However, as a number of astute observers have noted, altruistic instincts are actually fully consistent with "selfish gene" theories of evolutionary advantage. As Charles J. Lumsden and Edward O. Wilson put it in *Promethean Fire: Reflections on the Origin of Mind*,

> In its narrowest conception, evolution by natural selection—Darwinism—seems to imply survival of the fittest, the triumph of some individuals over others and the perpetuation of their genes in the next generation. But surrounding this unappealing image is a soft glow of altruistic behavior. We know that parents are willing to sacrifice a great deal on behalf of their children, even their own lives. Such behavior still conforms to Darwinism in the strict sense, because so long as children are preserved the parents' genes are passed on. If the self-sacrificing behavior of the parent is prescribed by some of the

favored genes, then that particular form of altruism will be spread through the population.[17]

The Selfish Biocosm hypothesis takes this analysis to a new level and suggests that broad-based altruism is, in effect, a cosmic necessity—a selectional constraint on those universes that will ultimately be capable of life-mediated reproduction. The selectional imperative of what might be provisionally called an "altruistic anthropic principle" derives from two logical requirements:

- A civilization capable of sustaining itself long enough to achieve the transcendent mental and technological capabilities needed to accomplish the feat of cosmological replication would seem to require a sustaining source of motivation more profound than that offered by short-term gratification of the material needs of the participants in that civilization.

- There is no apparent reason, apart from altruism, why a supremely advanced community of minds occupying the Olympian heights hypothesized to exist at the end of time (the eschaton era) would even bother to create a new baby universe they would never be privileged to inhabit.

The possibility of altruism thus appears to be, in the strictest evolutionary sense, a condition precedent to the process of cosmic replication that is the essence of the Selfish Biocosm hypothesis. Only a "selfish" biocosm that is, paradoxically, capable of the most profound act of altruism imaginable would satisfy the selectional constraint logically imposed by the altruistic anthropic principle.

This analysis is similar to that prompted by consideration of the most elusive and mysterious of the factors in the famous Drake equation ($N = Rf_p n_e f_l f_i f_c L$) discussed previously, which attempts to estimate the number of technological civilizations in our galaxy. Whereas many of the factors in the equation are relatively straightforward (the rate of formation of stars that are suitable for life, for instance, or the fraction of those stars with planetary systems), the factor *L*, which stands for the longevity of civilizations that have acquired the technological capacity for interstellar electromagnetic communication, is nearly impossible to quantify.

Michael Shermer, publisher of *Skeptic* magazine, opined in the August 2002 issue of *Scientific American* that *L* might be the crucial limiting variable, not only for purposes of SETI detection but also for the possibility of long-term survival of technologically advanced communities, either on Earth or elsewhere in the cosmos:

> I am an unalloyed enthusiast for the SETI program, but history tells us that civilizations may rise and fall in cycles too brief to allow enough to flourish at any one time to traverse (or communicate across) the vast and empty expanses between the stars. We evolved in small hunter-gatherer communities of 100 to 200 individuals; it may be that our species, and perhaps extraterrestrial species as well (assuming evolution operates in a like manner elsewhere), is simply not well equipped to survive for long periods in large populations.[18]

Only if our particular biocosm satisfies the daunting selectional constraint imposed by the cosmological requirement of broad-based altruism will Shermer's concerns ultimately be assuaged.

The Selfish Biocosm and Religion

A full treatment of the potential interaction of religious notions and the Selfish Biocosm hypothesis is beyond the scope of this book. However, it is important to remind ourselves that this process of interaction is likely to prove far more complex, interesting, and intellectually fruitful than proponents of science/religion apartheid would ever concede.

Far from being separate "nonoverlapping magisteria" that should be contemplated in isolation and separated by a rigid *cordon sanitaire* (as Stephen Jay Gould has implausibly suggested[19]), the overlapping domains of science, religion, and philosophy should be regarded as virtual rain forests of cross-pollinating ideas—precious reserves of endlessly fecund memes that are the raw ingredients of consciousness itself in all its diverse manifestations. The messy science/religion/philosophy interface should be treasured as an incredibly fruitful cornucopia of creative ideas—a constantly coevolving cultural triple helix of interacting ideas and beliefs that is, by far, the most precious of all the manifold treasures yielded by our history of cultural evolution on Earth.

In his classic Lowell Lectures delivered at Harvard in 1925, British philosopher Alfred North Whitehead offered a far more satisfying explanation of the rich interaction of scientific, religious, and philosophical thought than that proposed by Gould—and in the process put forward an intriguing explanation for the curious fact that European civilization alone had yielded the cultural phenomenon we know as scientific inquiry. Whitehead's theory was that "the faith in the possibility of science, generated antecedently to the development of modern scientific theory, is an unconscious derivative from medieval theology."[20] More specifically, he contended that

> the greatest contribution of medievalism to the formation of the scientific movement [was] the inexpugnable belief that every detailed occurrence can be correlated with its antecedents in a perfectly definite manner, exemplifying general principles. Without this belief the incredible labours of scientists would be without hope. It is this instinctive conviction, vividly poised before the imagination, which is the motive power of research—that there is a secret, a secret which can be unveiled.[21]

But precisely how, Whitehead asked, could this conviction have been so vividly implanted in the European mind? The answer, in Whitehead's view, lay not in the inherently obvious rationality of nature but rather in a peculiarly European habit of thought—a deeply ingrained, religiously derived, and essentially irrational faith in the existence of a rational natural order. He stated:

> When we compare this tone of thought in Europe with the attitude of other civilizations when left to themselves, there seems but one source for its origin. It must come from the medieval insistence on the rationality of God, conceived as with the personal energy of Jehovah and with the rationality of a Greek philosopher. Every detail was supervised and ordered: the search into nature could only result in the vindication of the faith in rationality.[22]

This habit of thought, Whitehead emphasized, consisted not of the "explicit beliefs of a few individuals" or a "mere creed of words" but rather an "instinctive tone of thought" resulting from "the impress on the European mind" of the "unquestioned faith of centuries."[23]

As Menas Kafatos and Robert Nadeau, two contemporary historians of science, have pointed out, this religiously inspired habit of thought played an indispensable role in the birth of the modern cosmological paradigm:

> The notion that the material world experienced by the senses is inferior to the immaterial world experienced by mind or spirit has been blamed for frustrating the progress of physics to at least the time of Galileo. In one very important respect, however, it made the first scientific revolution possible. Copernicus, Galileo, Kepler, and Newton firmly believed that the immaterial geometrical and mathematical ideas that "inform" or "give form to" physical reality had a prior existence in the mind of God and that doing physics was a form of communion with these ideas. It was this belief that allowed these figures to assume that geometrical and mathematical ideas could serve as transcriptions of the actual character of physical reality and could be used to predict the future of physical systems.[24]

Whitehead's analysis suggests that the historical interplay of these three great Western systems of thought and belief—Judeo-Christian religion, Greek philosophy, and Western science—can be viewed as a uniquely portentous instance of cultural coevolution and emergence that has powerfully shaped the path of intellectual history. Whitehead's rationale implies as well that a serious contemplation of the ontological implications of the Selfish Biocosm hypothesis could potentially advance that process of cultural coevolution in the spheres of religion, science, and philosophy simultaneously, with unpredictable but certainly interesting consequences in all three arenas of thought.

The Mind of Man → The Mind of God

Freeman Dyson has famously written that the idea of sufficiently evolved mind is indistinguishable from the mind of God. The Selfish Biocosm hypothesis takes Dyson's assertion of equivalence one step further by suggesting that there is a discernible and comprehensible evolutionary ladder by means of which mortal minds will one day ascend into the intellectual stratosphere that will be the domain of superminds—what Dyson would call the realm of God.

To use Hawking's terminology, the hypothesis implies that the mind of God is the natural culmination of the evolution of the mind of humans and other intelligent creatures throughout the universe, whose collective efforts conspire, admittedly without any deliberate intention, to effect a transformation of the cosmos from lifeless dust to vital, living matter capable of the ultimate feat of life-mediated cosmic reproduction. This act of transformation and replication, should it eventually succeed, will surely prove to be the defining conquest of ignorance by intelligence, the consummate victory of life over nonlife, the final triumph of mind over matter, and the ultimate revelation of the meaning of the cosmos.

To explore the surpassing wonder that is our cosmos is undoubtedly one of the greatest of intellectual pleasures and perhaps the most enlightening voyage of discovery that we shall ever experience as a species. As physicist John Wheeler once wrote,

> In the mind of every thinking person there is set aside a special room, a museum of wonders. Every time we enter that museum we find our attention gripped by marvel number one, this strange universe, in which we live and move and have our being. Like a strange botanic specimen newly arrived from a far corner of the earth, it appears at first sight so carefully cleaned of clues that we do not know which are the branches and which are the roots. Which end is up and which is down? Which part is nutrient-giving and which part is nutrient-receiving? Man? Or machinery?[25]

As we probe this strange specimen for clues as to its nature and origin, we hope desperately to glimpse not merely the details of its operation but some hint, however veiled, of its transcendent purpose. We seek, in short, to know and master the utility function of the universe, to summon out of the shadows a detailed foreknowledge of what Michael Polanyi called the "unthinkable consummation" that is the final step in the ongoing process of cosmic evolution and emergence.[26]

The great biologist Christian de Duve concluded his daring speculation on the meaning of life and the universe with these stirring words:

> If the universe is not meaningless, what is its meaning? For me, this meaning is to be found in the structure of the universe, which happens to be such as to produce thought by way of life and mind. Thought, in turn, is a faculty whereby the universe can reflect upon itself, discover its own structure, and apprehend such immanent entities as truth, beauty, goodness, and love. Such is the meaning of the universe, as I see it.[27]

The universe, de Duve would surely agree, cannot exhibit the miracle of conscious thought nor its finest manifestations—truth, beauty, goodness, and love—by itself. To do so, the universe urgently needs the assistance of humble mortals like ourselves.

As we go about our daily lives on planet Earth and occasionally pause to enter our own special museum of wonders, we might do well to remind ourselves of de Duve's profound reflection and of the deeper implications of the Selfish Biocosm hypothesis:

- that we participate, more or less unconsciously, in a cosmic life cycle of vast scope and unthinkable duration;

- that we are gradually and incrementally building an immense superstructure of matter and of mind that may one day become the very scaffolding of cosmic replication;

- that in our science and philosophy we instinctively reach for what is obviously far beyond our grasp, though hopefully not beyond the grasp of our ultimate progeny;

- that, like Isaac Newton, we are all humble seekers of enlightenment who occasionally chance to uncover a few pebbles of insight tossed our way from the great ocean of undiscovered truth that surges all around us;

- that we and all our fellow and future denizens of the vast cosmic biosphere are diligent masons who prepare, quite without conscious intent, the sturdy foundation for that singular moment in the far distant future—the rapturous instant of the eschaton—when minds born of mortals will become the mind of God.

There is, as Darwin once said, grandeur in this view of life.

Afterword

The Biocosm Beckons

The British philosopher, mathematician, and liberal humanist Bertrand Russell once expressed with soul-rending eloquence the existential angst entailed by a cosmology bereft of faith in divine direction and by the looming specter of the "night of nothingness" that confronts every thoughtful nonbeliever: "There is darkness without and when I die there will be darkness within. There is no splendor, no vastness, anywhere; only triviality for a moment, and then nothing."[1]

There is an equally disconcerting specter, of which Russell did not speak, that stems from a precisely contrary inference. Perceived vastness inheres unquestionably in the distant pageant of our evolutionary past. And there may be an equally palpable prospect of transcendent splendor in the distant future. However, at least some plausible incarnations of that looming splendor may prove to be utterly alien and bereft of any remnant of the human spirit.

The essence of the Selfish Biocosm hypothesis, after all, is that the anthropic qualities that our universe exhibits can be explained as incidental consequences of a cosmic replication cycle in which a cosmologically extended biosphere provides the means by which our cosmos duplicates itself and propagates one or more baby universes. The hypothesis suggests that the cosmos is "selfish" in the same sense that Richard Dawkins proposed that our genes are "selfish." Under the theory, the cosmos is "selfishly" intent upon the objective of replicating itself and self-replication is the hypothesized cosmic utility function. The emergence of life and intelligence are *means* to this end, not the *raison d'Lêtre* of the universe we inhabit. Indeed, the evolving cosmos is, under the theory, imbued with emergent properties of consciousness and intentionality that will eventually focus on its own set of objectives. These objectives may turn out to be disturbingly alien from a human perspective.

In other words, simply because the Selfish Biocosm hypothesis asserts that the emergence of life and ever more competent intelligence are highly robust phenomena, deeply inscribed in the basic laws of nature, it does not follow that the persistence or advancement of *human* life or intelligence or spirit is foreordained. If there is any practical lesson to be gleaned from the historical record of life on Earth, it is that evolution is a profligate wastrel—a careless spendthrift—spewing forth countless species that perish over the eons of geological time for every species that is privileged to survive or leave successor species as progeny. Why should the same not be true on a cosmic scale? Why should the catastrophe of entire worlds, as one prescient medieval philosopher put it, not be as commonplace in the vast universe as the casual extermination of a single species or even a single organism on planet Earth?

The realization that the Selfish Biocosm is indeed "selfish"—at least in a metaphorical sense—brings us full circle to the unforgiving vision of Jacques Monod: "The ancient covenant is in pieces: man knows at last that he is alone in the universe's unfeeling immensity, out of which he emerged only by chance. His destiny is nowhere spelled out; nor is his duty. The kingdom above or the darkness below: it is for him to choose."[2]

The Selfish Biocosm hypothesis suggests that it is indeed for us to choose—and that our choice has consequences for the ultimate question of whether mankind's progeny will participate in the construction of the cosmic future—whether our thoughts and those of our children will be woven into the fabric of what I have dared to call the mind of God. If we choose wisely—if, for instance, we manage to avoid the risk of nuclear holocaust or fatal degradation of the earthly environment and if we perceive and act upon a sense of moral obligation to future generations and to our fellow denizens of the cosmic biosphere, both here on Earth and possibly elsewhere—then we will have done all that we can possibly do to ensure that our distant progeny will be able to participate in the transcendent splendor that lies ahead.

There are no maps for us to follow in forging an optimal pathway to the future. And perhaps that is as it should be. Our plight, our challenge, and the looming adventure that awaits our species as we venture out into the vast wilderness of the undiscovered universe is summed up precisely and poignantly by an ancient Spanish proverb:

> Traveler, there are no roads. Roads are made by walking.

Appendix One

Genes Beget Memes and Memes Beget Genes: Modeling a New Catalytic Closure

By James Gardner

(As published in the May/June 1999 issue of Complexity, *the journal of the Santa Fe Institute)*

In a frequently quoted passage, Francis Crick [1] memorialized what he called the central dogma of molecular biology:

> The Central Dogma. This states that once information has passed into protein it cannot get out again. In more detail, the transfer of information from nucleic acid to nucleic acid, or from nucleic acid to protein may be possible, but transfer from protein to protein, or from protein to nucleic acid is impossible.

The central dogma undergirds classical Darwinian theory as well as contemporary restatements of it by Richard Dawkins and others. It sharply distinguishes evolutionary theory from Lamarckism (the notion that acquired characteristics can be inherited).

It is an article of faith among adherents to the central dogma that, in Dawkins' words, "the Lamarckian theory is completely wrong" [2]. However, to a degree that is largely unappreciated by orthodox theoretical biologists, the ongoing revolution in biotechnology renders the central dogma obsolete. The fact is that information can and does flow upstream into the genome from the particular extended phenotype we know as human civilization. As we enter what researcher J. Craig Venter calls "the century of biology" [3], a potent combination of scientific progress and an ever more frenetic entrepreneurial economy virtually ensures that the volume and velocity of this upstream informational flow will increase by orders of magnitude in the years ahead.

The advent of this historical phenomenon may, in the hindsight of history, prove to be a discontinuity event of enormous significance, perhaps equal in portent to the Cambrian explosion. It could represent what Dawkins [4] called in reference to the appearance of segmentation an "evolutionary watershed"—an event of macro-evolution which is primarily significant because it "open[s] floodgates to future evolution."[1]

The emerging capabilities of genetic engineering (GE) provide the technological pre-condition for explosively paced catalytic closure between the forces of genetic and cultural evolution. These new capabilities could conceivably create a radically new autocatalytic relationship between the two sets of forces which will enable the "old" replicators—Dawkins' selfish genes—and the "new" replicators—units of cultural transmission called culturgens or memes[2]—to coevolve in novel and unpredictable ways and at vastly enhanced speed.

Modeling this new relationship between the genetic and cultural fitness landscapes and mapping the potential pathways likely to be traced by GE-enabled coevolution of memes and genes may well turn out to be the signature challenge facing the nascent field of artificial life in the 21st century. The objective of this essay is to begin to measure the scope of this challenge and to suggest strategies for imposing a rudimentary investigatory structure upon it. My goal, in the words of Murray Gell-Mann [7], is to take a "crude look at the whole" of this new co-evolutionary phenomenon.

A New Catalytic Closure

In *The Origins of Order*, Stuart Kauffman [8] describes catalytic closure as the threshold at which a sufficiently complex set of complex polymers becomes an autocatalytic set, whose connectivity requirements he defines as follows:

> The connectivity requirements allowing an autocatalytic set of polymers to exist are simply stated. Each member of the set must have its formation catalyzed by at least one member of the set. Furthermore,

> there must be connected catalysis pathways leading from a maintained exogenous food set to all members of the autocatalytic set.

Kauffman has recognized the potential applicability of autocatalytic modeling techniques to the social sciences:

> The largest intellectual agenda of this chapter is based on the presumption that—by analyzing a variety of grammars from regions of a parameterized grammar space, each grammar a kind of hypothetical set of laws of chemistry or functional complementarity—a few broad regimes will emerge. Where we can map such generic behaviors onto molecular, organismic, neural, psychological, economic or cultural data, we may have found the functional universality class needed to explain phenomena in these areas of chemistry, biology, and the social sciences.[3]

The present essay proceeds from a more modest premise: that a special class of cultural advances—progress in genetic engineering—will likely yield changes in the human genotype by application of germline genomic therapy and that these changes will yield subsequent cultural changes which will, in turn, be manifested in the form of altered cultural proclivities regarding further bioengineering of the human genome. This feedback loop may create a classic autocatalytic set where the catalysts are not prebiotic polymers but rather coevolving memes and genes.[4]

Deciphering the possible catalytic pathways likely to be traced by the ensuing meme/gene coevolutionary process is the daunting task that lies ahead. An appropriate way to begin to illuminate the scope of this challenge is to pose a series of questions aimed at elucidating key objectives of a potential research program.

Deciphering the Utility Function of Gene/Meme Coevolution

The notion that human culture and the human genome have co-evolved over the deep history of our species is not new. Indeed, the basic concept of human gene/culture coevolution is gradually gaining acceptance in the scientific community, while continuing to generate considerable controversy. Edward O. Wilson [6] recently summarized the emerging consensus as follows:

> What, in final analysis, joins the deep, mostly genetic history of the species as a whole to the more recent cultural histories of its far-flung societies? . . . It can be stated as a problem to be solved, the central problem of the social sciences and the humanities, and simultaneously one of the great remaining problems of the natural sciences. . . . A few researchers, and I am one of them, even think they know the approximate form the answer will take. From diverse vantage points in biology, psychology and anthropology, they have conceived a process called gene-culture coevolution. In essence, the conception observes, first, that to genetic evolution the human lineage has added the parallel track of cultural evolution, and, second, that the two forms of evolution are linked. I believe the majority of contributors to the theory during the past twenty years would agree to the following outline of its principles:
>
> Culture is created by the communal mind, and each mind in turn is the product of the genetically structured human brain. Genes and culture are therefore inseverably linked. But the linkage is flexible, to a degree still mostly unmeasured. The linkage is also tortuous: Genes prescribe epigenetic rules, which are the neural pathways and regularities in mental development by which the individual mind assembles itself. The mind grows from birth to death by absorbing parts of the existing culture available to it, with selections guided by the epigenetic rules inherited by the individual brain.[5]

Human cultural evolution, Wilson points out, has also shaped profoundly the fitness landscape within which this coevolutionary process occurs:

> What is truly unique about human evolution, as opposed say to chimpanzee or wolf evolution, is that a large part of the environment shaping it has been cultural. Therefore, construction of a special environment is what culture does to the behavioral genes. Members of past generations who used their culture to best advantage, like foragers gleaning food from a surrounding forest, enjoyed the greatest Darwinian advantage. During prehistory their genes multiplied, changing brain circuitry and behavior traits bit by bit to construct human nature as it exists today.[6]

What is not clear is the extent to which cultural evolution, at least over the short to medium term, furthers the agenda of the "selfish gene." As Wilson puts it:

> Particular features of culture have sometimes emerged that reduce Darwinian fitness, at least for a time. Culture can indeed run wild for a while, and even destroy the individuals that foster it.[7]

Dawkins [10] makes the same point more systematically, contending that the fitness of meme-replicators is not invariably reducible to their ability to improve gene survival:

> Fundamentally, the reason why it is good policy for us to try to explain biological phenomena in terms of gene advantage is that genes are replicators. As soon as the primeval soup provided conditions in which molecules could make copies of themselves, the replicators themselves took over. For more than three thousand million years, DNA has been the only replicator worth talking about in the world. But it does not necessarily hold these monopoly rights for all time. Whenever conditions arise in which a new kind of replicator *can*

> make copies of itself, the new replicators *will* tend to take over, and start a new kind of evolution of their own. Once this new evolution begins, it will in no necessary sense be subservient to the old. The old gene-selected evolution, by making brains, provided the "soup" in which the first memes arose. Once self-copying memes had arisen, their own, much faster, kind of evolution took off. We biologists have assimilated the idea of genetic evolution so deeply that we tend to forget that it is only one of many possible kinds of evolution.

These observations by Dawkins and Wilson underscore the first set of difficult questions that arise in the context of the present inquiry: what can we learn about the precise nature of the utility function served by GE-enabled gene/meme coevolution? While so-called ultra-Darwinists might contend that DNA survival is the ineluctable utility function of all organic life forms (including humanity), this assertion overlooks the possibility implicitly acknowledged by Wilson and Dawkins that "selfish memes" may increasingly be calling the tune in life's complex symphony. As Dawkins notes expressly, there is no reason to assume that the utility function of memes in a memetic fitness landscape is completely reducible to their capacity to affect favorably the transmission of genetic information from one generation into the next. It is the memes' *own* survivability that matters, at least in part, in the realm of memetic evolution.[8]

The onset of GE-enabled gene/meme coevolution introduces a new level of complexity to the analysis. Memes evolve within—and continuously modify—a complex informational ecology that we may broadly refer to as the cultural environment or infosphere. This specialized information-rich environment and the poorly understood processes which give it shape and substance relate to the underlying genetic fitness landscape in complicated and often counter-intuitive ways. Moreover, the nature of the relationship of these two overlapping fitness landscapes is dynamic and subject to rapid, non-linear change as, for example, progress in the science of biotechnology (which represents at least one of the leading edges of memetic evolution) gives humanity the ability to directly modify the DNA code governing genetic transmission.

Some basic issues deserving of analysis and investigation in the general context of the task of deciphering the gene/meme coevolutionary utility function are:

- Since "[c]ultural evolution operates many orders of magnitude faster than genetic evolution"[9] does the hypothesized autocatalytic loop imply a massive acceleration in the rate of genetic change?

- To what extent should the creation of this hypothetical feedback loop (from gene to meme and back to gene through deliberate biotechnological manipulation) be characterized as the moment of triumph of the meme—a crucial inflection point in life's long pageant when cultural forces have fundamentally displaced genetic variation and selection as the principal engine of evolutionary change? What metric can be employed to measure the degree of displacement? Over what time scale should the putative displacement be assessed?

- If memes have indeed supplanted the "selfish genes" of Darwinian evolution as the principal engines driving genetic modification and adaptation, then what novel pathways might gene/meme coevolution take in the decades and centuries ahead? Do we stand on the brink of a macro-evolutionary threshold roughly analogous to the Cambrian explosion [12]? Or is the premise of genetic conservatism favored by Wilson[10] likely to prove accurate, at least in the short to medium term? If so, what time horizons are likely to limit the applicability of this premise? What factors—technological and/or social—will tend to extend or contract such time horizons?

- Can the modeling techniques under development in the nascent field of artificial life be employed to map these alternative futures and evaluate their probabilities? If not currently capable of such modeling, what technical improvements are necessary to give a-life modeling tools this capability?

- Could these alternative futures be deliberately manipulated using techniques that might be referred to collectively as memetic engineering [13]?
- To what extent is our human capacity for analysis and manipulation of this coevolutionary process inherently bounded inasmuch as our investigatory tools (including our mental capacities) are themselves artifacts of the process under investigation?
- Finally, to what extent is a meme-dominated coevolutionary process likely to become "platform neutral" and eventually independent of the organic phenotypes in which the phenomenon of memetic evolution first arose?

A Methodological Strategy

Can the daunting research endeavor intimated by the queries listed above be undertaken credibly or should questions like these be consigned to the province of philosophy, politics and religion? I believe that a carefully designed research agenda can begin to address these difficult questions, at least in a modest way.

My suggested approach is to begin with a set of possible cultural pathways prefigured by speculations in the popular and scientific media. These media, after all, comprise collectively a kind of memetic "primeval soup"[11] in which cultural evolution both occurs and is further catalyzed. In particular, the popular media (including preeminently television) dramatically impact public attitudes toward health and wellness topics [14], which are likely to be among the principal memetic or cultural attractors shaping the putative GE-enabled meme/gene coevolutionary process.[12]

After describing briefly three key sets of cultural attractors, I will outline a possible investigatory approach employing agent-based software tools to model GE-enabled meme/gene coevolutionary processes likely to be catalyzed by the interaction of these attractors and the availability of ever more sophisticated GE techniques.

Cultural Attractor Set #1: Disease/Disability Avoidance

Even Edward Wilson, who predicts that "future generations will be genetically conservative" and "will resist hereditary change" in order to "save the emotions and epigenetic rules of mental development because these elements compose the physical soul of the species," concedes that "repair of disabling defects" through human germline engineering is probable.[13] This conclusion finds support in a study by the Office of Technology Assessment (now defunct) which in its evaluation of the Human Genome Project stated that "one of the strongest arguments for supporting human genome projects is that they will provide knowledge about the determinants of the human condition" and the diseases and debilitating conditions "that are at the root of many current societal problems" [15].

A recent paper by Gregory Stock and John Campbell [16] indicates that the technical capability to utilize germline engineering to achieve such therapeutic objectives "may be much nearer than many imagine":

> Recent work on human artificial chromosomes suggests it may soon be possible to reliably insert gene cassettes into them for injection into cells, including embryonic stem cells and eggs. A look at the accelerated timetable for completing the human genome project and the massive energy being directed towards developing somatic therapies and understanding genetic regulation indicates that gene insertions to confer resistance to AIDS, cancer, or even some aspects

> of aging itself may be imminent. Germline engineering may open up entirely new approaches to therapy.

Clearly, then, the broad objectives of disease and disability avoidance in our progeny would seem to comprise an important set of cultural attractors which should be incorporated in any realistic attempt to model GE-enabled meme/gene coevolution.

Cultural Attractor Set #2: Enhancement of Human Capabilities

The next broad set of cultural attractors that would appear to qualify for incorporation in the proposed coevolutionary model may be labeled enhancement of human capabilities. Wilson offers a useful summary of the potential scope of this set of attractors:

> [W]hat about altering genes in order to enhance mathematical and verbal ability? To acquire perfect pitch? Athletic talent? Heterosexuality? Adaptability to cyberspace? In a wholly different dimension, citizens of states and then of all humanity might choose to make themselves less variable, in order to increase compatibility. Or the reverse: They might choose to diversify in talent and temperament, aiming for varied personal excellence and thus the creation of communities of specialists able to work together at higher levels of productivity. Above all, they will certainly aim for greater longevity.[14]

Many ethicists have cautioned against premature pursuit of these cultural objectives through GE techniques. Michael J. Riess and Roger Straughan [17], for instance, advocate a ban on germline manipulation aimed at enhancement (but not fault correction):

> Despite the difficulties . . . of distinguishing in all cases genetic engineering to correct faults (such as cystic fibrosis, hemophilia or cancers) from genetic engineering to enhance traits (such as intelligence, creativity, athletic prowess or musical ability), the best way forward may

> be to ban germline therapy intended only to enhance traits, at least until many years of informed debate have taken place.

The fact that this particular cultural attractor set is fraught with ethical, legal, social, and even spiritual issues does not disqualify it from incorporation in the proposed coevolutionary model. On the contrary, this feature makes rigorous analysis all the more important. One fundamental objective of the analysis should be to assess the strength and durability of cultural barriers impeding germline therapies aimed at enhancement of desirable human traits and thus to seek an answer to the momentous question posed by Edward O. Wilson in *On Human Nature* [18]:

> The human species can change its own nature. What will it choose? Will it remain the same, teetering on jerrybuilt foundation of partly obsolete Ice-Age adaptations? Or will it press on toward still higher intelligence and creativity, accompanied by a greater—or lesser—capacity for emotional response? New patterns of sociality could be installed in bits and pieces. It might be possible to imitate genetically the more perfect nuclear family of the white-handed gibbon or the harmonious sisterhoods of the honeybees. *But we are talking here about the very essence of humanity. Perhaps there is something already present in our nature that will prevent us from ever making such changes.*[15]

Cultural Attractor Set #3:
Cosmetic Germline Engineering

Closely related to the preceding attractor set is germline engineering aimed at cosmetic objectives. *The New York Times* [19] reported recently on a veritable explosion in cosmetic surgery, including a boom in plastic surgery for adolescents:

> At least 14,000 adolescents nationwide had cosmetic surgery in 1996, a slight increase from 1992, when the boom began, according to data from the American Society of Plastic and Reconstructive Surgeons.

> In all age groups, 700,000 procedures were done last year, up 70 percent in four years. But professionals agree that those numbers are a vast understatement, perhaps by as much as half, since they do not include the many procedures now done by dermatologist, ophthalmologists, ear, nose and throat specialists, dentists and others.

As the *Times* article notes, this phenomenon is arguably driven, at least in part, by invidious comparisons by television viewers between their own bodily features and those of popular cultural icons like "'Baywatch' babes and Victoria's Secret models." If such ideals of physical attractiveness are sufficiently powerful to stimulate rapid increases in the rate of plastic surgery, might they also influence the behavior of prospective parents browsing in what Robert Nozick has aptly termed the "genetic supermarket"?[16]

The three sets of cultural attractors described above are not intended to be exhaustive. However, they do capture some of the key cultural forces identified to date in both popular and scientific commentary concerning the ethical, legal and social implications of genetic engineering. As such, they would seem to qualify for incorporation in an initial agent-based modeling system aimed at tracing potential pathways of GE-enabled meme/gene coevolution.

Selection of Modeling Strategies

In an article describing a December, 1997, conference convened by the Culture Group at the Santa Fe Institute, Henry Wright [20] expressed the view of conference participants that there was "value of continuing to model with a diversity of platforms" including Swarm and models inspired by the Brookings Sugarscape model. These and other agent-based models (including the synthetic consumer model developed by the Cooper & Lybrand Emergent Solutions Group [21]) offer potentially useful platforms for modeling the possible GE-enabled meme/gene coevolutionary process hypothesized in this essay.

Regardless of the platform utilized for initial experimentation, a useful first step would be to conduct an in-depth benchmark poll assessing public attitudes toward the particular cultural attractors selected for inclusion in the model. Information gleaned from such a poll could form the initial data set from which agent characteristics could be extracted, including demographic and psychographic data.

The goal of the modeling should be to attempt to achieve, in the particular context of GE-enabled meme/gene coevolution, the general objective articulated by Joshua Epstein and Robert Axtell in *Growing Artificial Societies: Social Science from the Bottom Up* [22]:

> The broad aim of this research is to begin the development of a more unified social science, one that embeds evolutionary processes in a computational environment that simulates demographics, the transmission of culture, conflict, economics, disease, the emergence of groups, and coadaptation with an environment, all from the bottom up.

If even moderately successful, such modeling could form a useful component of the ongoing effort undertaken in conjunction with the Human Genome Project ("HGP") to assess the ethical, legal and social implications of genetic research and genetic engineering.[17]

The Ethical, Legal and Social Implications ("ELSI") research initiative administered by National Institutes of Health ("NIH") and the Department of Energy ("DOE") is viewed by administrators responsible for funding HGP research as an essential element of the HGP endeavor. As stated in a recent article in *Science* [23]:

> [T]he NIH and DOE are acutely aware that advances in the understanding of human genetics and genomics will have important implications for individuals and society. Examination of the ethical, legal, and social implications of genome research is, therefore, an integral and essential component of the HGP.

The ELSI initiative has established major goals for the next five years which include "explor[ation of] ways in which new genetic knowledge may interact with a variety of philosophical, theological, and ethical perspectives." ELSI administrators acknowledge that achieving this and other key 5-year ELSI research goals will require collaboration between "biological and social scientists, health care professionals, historians, legal scholars, and others . . . committed to exploration of these issues as the project proceeds."

Utilization of the agent-based modeling approach advocated in this essay could conceivably add a new level of rigor to research conducted under the aegis of the ELSI goal quoted above.

Conclusion

In *The Origin of Species* Charles Darwin masterfully deployed the art of metaphor to elucidate his central hypothesis. Artificial selection was the primary intellectual model that guided Darwin in his epic quest to solve the mystery of the origin of species and demonstrate in principle the plausibility of his theory that variation and natural selection were the prime movers responsible for the phenomenon of speciation.

There is a huge and largely unappreciated puzzle embedded in Darwin's use of this metaphor, which may be summarized as follows: how, precisely, do the cultural norms which guide the process of artificial selection themselves arise and how do they—consciously, unconsciously or both—shape that process? Moreover, how does the phenomenon of culture/phenotype convergence—itself the fruit of this coevolutionary process—feed back into the dynamics of the process?

These are the questions, largely unanswered by theoretical biologists, which are posed with new urgency by the contemporary phenomenon of GE-enabled meme/gene coevolution. They deserve careful and relentless investigation as we enter the century of biology.

Note: Superscripts 1–17 in appendix 1 refer to the following notes. Bracketed numbers 1–23 correspond with the references listed on page 250.

Notes

1. See Dawkins [4], p. 218.
2. The term "culturgen" was originally favored by Edward Wilson, but he recently acknowledged the ubiquity of the term "meme." Compare Lumsden and Wilson [5] with Wilson [6].
3. See Kauffman [8], p. 404.
4. For one set of views about the future direction of GE-enabled meme/gene coevolution, see the speculations of Lee Silvers [9] about the possible divergence of humanity into separate species, which he refers to as the "GenRich" and the "Naturals."
5. See Wilson [6], pp. 137–138.
6. See Wilson [6], p. 180.
7. See Wilson [6], p. 171.
8. See Dawkins [10], p. 200 ("What we have not previously considered is that a cultural trait may have evolved in the way that it has, simply because it is *advantageous to itself*"). See also Dennett [11] ["The most important point Dawkins makes, then, is that there is no *necessary* connection between a meme's replicative power, its 'fitness' from *its* point of view, and its contribution to our fitness (by whatever standard we judge that)"].
9. See Dennett [11], p. 339.
10. See Wilson [6], p. 303.
11. See Dawkins [10], p. 192.
12. A memetic or cultural attractor is, for purposes of this analysis, a set of cultural points or states in cultural state space to which cultural trajectories within some volume of cultural state space converge asymptotically over time. Compare this with Kauffman [8], p. 177. A similar concept was articulated by Dennett [11], who uses the nomenclature "cranes of culture" (pp. 355 et seq).
13. See Wilson [6], p. 103.
14. See Wilson [6], p. 301.
15. Emphasis added.
16. See Reiss and Straughan [17], p. 220.
17. For a general description of the breadth and range of the issues that will be addressed as a part of this initiative (dubbed "ELSI"), consult the following website: http://www.nhgri.nih.gov/98plan/elsi/.

References

1. Crick F. "On Protein Synthesis." *Symp. Soc. Exp. Biol.* 12, 1958, 138–167.
2. Dawkins R. *The Extended Phenotype.* Oxford, New York, 1989, 289.
3. "The Century of Biology." *Business Week.* August 31, 1998, 86.
4. Dawkins R. "The Evolution of Evolvability." In: *Artificial Life, Volume VI.* Santa Fe Institute, Santa Fe , NM, 217–218.
5. Lumsden C. J.; Wilson E.O. *Genes, Mind, and Culture.* Harvard University Press, Cambridge, MA. 1981.
6. Wilson E.O. *Consilience.* Vintage, New York, 1998, 148.
7. Gell-Mann M. *The Quark and the Jaguar.* W. H. Freeman, New York, 1994, 346.
8. Kauffman S. A. *The Origins of Order.* Oxford, New York, 1993, 309.
9. Silvers L. M. *Remaking Eden.* Avon, New York, 1997, 4–7.
10. Dawkins R. *The Selfish Gene.* Oxford, New York, 1989, 193–194.
11. Dennett D. C. *Darwin's Dangerous Idea.* Touchstone, New York, 1995, 363.
12. Gould S. J. *Wonderful Life.* Norton, New York, 1990.
13. Gardner J. N. "Memetic Engineering." *Wired.* 4.05, 1996, 101.
14. National Health Council "Americans Talk about Science and Medical News: The National Health Council Report." Roper Starch Worldwide, Inc., 1997.
15. U.S. Congress, Office of Technology Assessment: *Mapping Our Genes.* U.S. Government Printing Office, Washington D.C., 1988, 85.
16. Stock G.; Campbell J. "Human Germline Engineering. The Prospects for Commercial Development." Electronic workshop presentation on "Advanced Transgenesis and Cloning: Genetic Manipulation in Animals." ATC Workshop paper No. 18.
17. Reiss M. J.; Straughan R. *Improving Nature? The Science and Ethics of Genetic Engineering.* Cambridge University Press, Cambridge, UK, 1996, 220.
18. Wilson E. O. *On Human Nature.* Harvard University Press, Cambridge, MA, 1978, 208.
19. Gross J. "In Quest for the Perfect Look, More Girls Choose the Scalpel." *The New York Times* (on-line edition), November 29, 1998.
20. Wright H. "Culture Group Meeting. Agent-Based Modeling of Small-Scale Societies." *SFI Bulletin* 13, 1998, 21.
21. Farrell W. *How Hits Happen.* HarperBusiness, New York, 1998.
22. Epstein J.; Axtell R. *Growing Artificial Societies. Social Science From the Bottom Up.* Brookings Institution, Washington, D.C., 1995, 19.
23. Collins, et al. "New Goals for the U.S. Human Genome Project. 1998–2003." *Science.* 282, 682.

Appendix Two

Assessing the Computational Potential of the Eschaton: Testing the Selfish Biocosm Hypothesis

By James N. Gardner

(As published in the July/August 2002 issue of the Journal of the British Interplanetary Society)

Abstract

The Selfish Biocosm (SB) hypothesis asserts that the anthropic qualities which our universe exhibits can be explained as incidental consequences of a cosmic replication cycle in which a cosmologically extended biosphere supplies two of the essential elements of self-replication identified by von Neumann. It was previously suggested that the hypothesis implies (1) that the emergence of life and intelligence are key epigenetic thresholds in the cosmic replication cycle, strongly favored by the physical laws and constants which prevail in our particular universe and (2) that a falsifiable implication of the hypothesis is that the emergence of increasingly intelligent life is a robust phenomenon, strongly favored by the natural processes of evolution which result from the interplay of those laws and constants. Here I propose a further falsifiable implication of the SB hypothesis: that there exists a plausible final state of the cosmos which exhibits maximal computational potential. This predicted final state—the Omega Point or eschaton—appears to be not inconsistent with Lloyd's description of the ultimate computational device: a computer as powerful as the laws of physics will allow.

Keywords: Anthropic, biocosm, computation, cosmology, ekpyrotic

1. Introduction

In a recent paper [1], I first advanced the hypothesis that the anthropic qualities which our universe exhibits can be explained as incidental consequences of a cosmic replication cycle in which the emergence of a cosmologically extended biosphere could conceivably supply two of the logically essential elements of self-replication identified by von Neumann in [2]: a controller and a duplicating device. The hypothesis advanced in [1] was an attempt to extend and refine Smolin's conjecture [3] that the majority of the anthropic qualities of the universe can be explained as incidental consequences of a process of cosmological replication and natural selection (CNS) whose utility function is black hole maximization. Smolin's conjecture differs crucially from the concept of eternal chaotic inflation advanced by Linde [4] in that it predicts a cosmological evolutionary process with a specific and discernible utility function. It is this aspect of Smolin's conjecture rather than the specific utility function he advocates that renders his theoretical approach genuinely novel.

As noted previously [5,6], Smolin's conjecture suffers from two evident defects: (1) the fundamental physical laws and constants do not, in fact, appear to be fine-tuned to favor black hole maximization and (2) no mechanism is proposed corresponding to two logically required elements of any von Neumann self-replicating automaton: a controller and a duplicator.

Theories of cosmological eschatology previously articulated in [7,8,9] predict that the ongoing process of biological and technological evolution is sufficiently robust and unbounded that, in the far distant future, a cosmologically extended biosphere could conceivably exert a global influence on the physical state of the cosmos. A related set of insights from complexity theory indicates that the process of emergence resulting from such evolution is essentially unbounded.

A synthesis of these two sets of insights yields the two key elements of the Selfish Biocosm (SB) hypothesis. The essence of that synthesis is that the ongoing process of biological and technological evolution and emergence could conceivably function as a von Neumann controller and that a cosmologically extended biosphere could, in the very distant future, function as a von Neumann duplicator in a process of cosmological replication.

In [10], it was suggested that a falsifiable implication of the SB hypothesis is that the process of the progression of the cosmos through critical epigenetic thresholds in its life cycle, while perhaps not strictly inevitable, is relatively robust. One such critical threshold is the emergence of human-level and higher intelligence, which is essential to the eventual scaling up of biological and technological processes to the stage at which those processes could conceivably exert a global influence on the state of the cosmos. Four specific tests of the robustness of the emergence of human-level and higher intelligence were proposed in [10].

One of tests proposed in [10] was that the ongoing SETI research endeavor will eventually succeed in detecting evidence of extraterrestrial intelligence (ETI). Some would argue that Enrico Fermi's paradox concerning the observed absence of extraterrestrials on Earth ("Where are they?" Fermi famously asked) has already falsified the robustness claim of the SB hypothesis. However, there are other explanations than nonexistence that may explain the absence to date of evidence of ETI. For useful reviews of the issues raised by Fermi's paradox and other SETI-related issues, see [11] and [12].

The tests proposed in [10] do not address the issue of whether there exists a plausible final state of the cosmos which is both consistent with the best astrophysical evidence currently available and which exhibits maximal computational potential. The plausible existence of such a final state—which may be referred to as the eschaton or Omega Point—is a falsifiable prediction of the SB hypothesis.

2. Predicted Characteristics of the Eschaton

What characteristics must the eschaton exhibit in order to function as a von Neumann duplicator? The performance requirements are exacting indeed. If, as previously suggested [10], the laws and constants of nature which prevail in our universe are functionally analogous in the SB scenario to the genetic code in an earthly organism, then the eschaton must possess the capability to both (1) replicate the "cosmic code" [13] and (2) reliably transmit it to a new "baby" universe [14]. While there has been limited speculation about the actual technology by which cosmic code replication and transmission might be accomplished (i.e., Calabi-Yau shape engineering) [1], the important point for purposes of this paper is that there appears to be a tentative consensus that any variety of final-state cosmological engineering capable of exerting a global influence on the state of the cosmos will require maximal computational capability [15, 16]. This, then, is the hypothesized threshold characteristic which the eschaton must possess in order to be capable of functioning as a von Neumann duplicator in the context of the SB hypothesis: maximal computational capability.

3. The Modified Ekpyrotic Cyclic Universe Scenario: A Plausible Cosmic Eschatology

The next question is whether there exists a plausible cosmological eschatology which could yield an eschaton with maximal computational capability. A preliminary issue is whether there are candidate physical eschatology scenarios available which would physically support such a capability. Of particular interest for purposes of this paper is a suggested modification and simplification of the recently proposed ekpyrotic cosmological scenario [17] that provides a framework for modeling a cyclic universe [18,19]. In this model "the Universe undergoes an endless sequence of cosmic epochs which begin with the Universe expanding from a 'big bang' and end with the Universe contracting to a 'big crunch.'" [19] The "linchpin of the cyclic picture is safe passage through the cosmic singularity, the transition from the big crunch to big bang." [19] String theory provides the basis for the hypothesized safe passage by offering the possibility that "what appears to be a big crunch in the 4d effective theory actually corresponds to the momentary collapse of an additional fifth dimension." [19] This provides an alternative to the scenario in "ordinary 4d general relativity" wherein "the big crunch is interpreted as the collapse and disappearance of four-dimensional space-time" and "there is no sign that a transition is possible." [19] By contrast, under the modified ekpyrotic cyclic universe scenario "temperature and density are finite as one approaches the crunch" and "there is nothing to suggest that time comes to an end when the fifth spatial dimension collapses"; on the contrary, the "most natural possibility is that time continues smoothly" across the Big Crunch/Big Bang threshold. [19] While the characteristics of the Big Crunch/Big Bang transition have not yet been modeled rigorously in string theory, it is sufficient for purposes of this paper to note that the ekpyrotic cyclic universe scenario represents a plausible cosmological eschatology, derived from string theory and consistent with the latest observational data. Under this theory, the eschaton would correspond to the physical state of the cosmos immediately prior to and during the Big Crunch.

It must be conceded that the modified ekpyrotic cyclic universe scenario is a relatively new conceptual entrant in the rapidly inflating field of string theory–inspired cosmological conjecture. However, the scenario has already demonstrated that it possesses key advantages over alternative scenarios, including its incorporation of the currently observed phenomenon of cosmic acceleration. As stated in [19]:

> What is the role of dark energy and the current cosmic acceleration [in the modified ekpyrotic cyclic universe scenario]? Clearly, dark energy and the current cosmic acceleration play an essential role in the cyclic model both by reducing the entropy and black hole density of the previous cycle, and triggering the turnaround from an expanding to a contracting phase. (In all other cosmologies to date, including inflation, dark energy has no essential role.)

4. Computational Capacity of the Eschaton Under the Modified Ekpyrotic Cyclic Scenario

The crucial question is whether—and, in principle, how capably—the eschaton yielded by this theoretical scenario can compute. The canonical assumption has been that any scenario pursuant to which the universe "ends in fire" (i.e., a Big Crunch) precludes the possibility of the persistence of information processing (or life, for that matter) through the Big Crunch era. While the modified ekpyrotic cyclic universe scenario sketched out above casts some doubt on this conventional wisdom in and of itself (by eliminating the necessity for a 4d singularity at the instant of the Big Crunch), the most powerful indication that this scenario requires a rethinking of the canonical assumption comes from a branch of science far removed from cosmology: computational theory.

In [20] Lloyd sought to explore the ultimate physical limits of computation as determined by the speed of light, the quantum scale and the gravitational constant. The "ultimate computer" described by Lloyd's exercise is a computing device as powerful as the laws of physics will allow. The "typical state" of the memory of this hypothesized computing device "looks like a plasma at a billion degrees Kelvin—like a thermonuclear explosion or a little piece of the Big Bang." [20] This is because "to take full advantage of the memory space available" the ultimate computing device "must turn all its matter into energy." [20]

If Lloyd's conclusions are correct, they indicate that the characteristics of the eschaton under the modified ekpyrotic cyclic universe scenario—a non-singular condition characterized by extremely high temperature and density—are not inconsistent with the characteristics of an ultimately powerful computational device. In short, under this scenario the eschaton could plausibly constitute a computer as powerful as the laws of physics will allow.

5. Conclusion

The SB hypothesis, while obviously speculative, can qualify as a genuine scientific hypothesis (as opposed to a metaphysical speculation) only if it yields falsifiable predictions. One set of such predictions was provided in [10]. This paper proposes a second major falsifiable prediction of the SB hypothesis: that there exists a plausible final state of the cosmos which exhibits maximal computational potential. This predicted final state—the eschaton—appears, under the modified ekpyrotic cyclic universe scenario, to be not inconsistent with Lloyd's description of the physical attributes of the ultimate computational device: a computer as powerful as the laws of physics will allow. It will be interesting to see whether further refinements to the cyclic universe scenario and Lloyd's characterization of the ultimate physical limits of computation remain consistent with the SB hypothesis.

References

1. J. Gardner, "The Selfish Biocosm: Complexity as Cosmology", *Complexity*, 5, No. 3, pp. 34–45, 2000.
2. J. von Neumann, "On the General and Logical Theory of Automata", 1948.
3. L. Smolin, *"The Life of the Cosmos"*, Oxford University Press 1997.
4. A. Linde, "The Self-Reproducing Inflationary Universe", in *Scientific American*, 9(20), pp. 98–104, 1998.
5. M. Rees, *"Before the Beginning"*, Addison-Wesley, 1997.
6. J. Baez, on-line commentary on *"The Life of the Cosmos"*, (available at: http://www.aleph.se/Trans/Global/Omega/smolin/txt.
7. R. Kurzweil, *"The Age of Spiritual Machines"*, Viking, 1999.
8. J. Wheeler, *"At Home in the Universe"*, AIP Press, 1994.
9. F. Dyson, *"Infinite in All Directions"*, Harper, 1988.
10. J. Gardner, "Assessing the Robustness of the Emergence of Intelligence: Testing the Selfish Biocosm Hypothesis", *Acta Astronautica*, 48, No. 5–12, pp. 951–955, 2001.
11. S. J. Dick, *"The Biological Universe"*, Cambridge University Press, 1996.

12. M. Hart and B. Zuckerman, Editors, *"Extraterrestrials: Where Are They?"*, Pergamon Press, 1982.

13. H. Pagels, *"The Cosmic Code"*, Bantam, 1983.

14. R. Dawkins, "Implications of Natural Selection and The Laws of Physics", Lee Smolin/Richard Dawkins on-line colloquy available at: http://www.edge.org/discourse/index/cgi?OPTION=VIEW&THREAD=richard-dawkins/?smolin; p. 3.

15. J. Barrow and F. Tipler, *"The Anthropic Cosmological Principle"*, Oxford University Press, 1988.

16. F. Tipler, *"The Physics of Immortality"*, Anchor Books/Doubleday, 1995.

17. J. Khoury, B. A. Ovrut, P. J. Steinhardt and N. Turok, "The Ekpyrotic Universe: Colliding Branes and the Origin of the Hot Big Bang", hep-th/0103239, *Phys. Rev. D*, in press, 2001.

18. J. Khoury, B. A. Ovrut, N. Seiberg, P. Steinhardt and N. Turok, "From Big Crunch to Big Bang", hep-th/0108187, 2001.

19. P. Steinhardt and N. Turok, "Cosmic Evolution in a Cyclic Universe", hep-th/0111098, 2001.

20. S. Lloyd, "Ultimate physical limits to computation", *Nature*, 406, pp. 1047–1054, 2000.

Notes

Introduction

1. Freeman J. Dyson, "Science & Religion: No Ends in Sight,"" *The New York Review of Books*, 28 March 2002, 6.
2. Richard Dawkins, *The Selfish Gene* (Oxford: Oxford University Press, 1976).
3. J. Richard Gott III, *Time Travel in Einstein's Universe* (New York: Houghton Mifflin, 2001).
4. Kenneth R. Miller, *Finding Darwin's God* (New York: Cliff Street Books, 2000), 130.
5. Thomas Gold, *The Deep Hot Biosphere* (New York: Copernicus, 2001), 37.
6. Michael Shermer, *The Borderlands of Science* (Oxford: Oxford University Press, 2001).
7. Thomas S. Kuhn, *The Structure of Scientific Revolutions* (Chicago: University of Chicago Press, 1970).
8. James Gleick, *Chaos: Making a New Science* (New York: Viking Penguin, 1988), 37.
9. John L. Casti, *Would-Be Worlds* (New York: Wiley, 1997).
10. Ibid., xi.
11. For a detailed discussion of the concept of consilience, see Edward O. Wilson, *Consilience* (New York: Knopf, 1998).
12. James N. Gardner, "Mastering Chaos at History's Frontier: The Geopolitics of Complexity,"" *Complexity* 3, no. 2 (1997): 28–32.
13. James N. Gardner, "Genes Beget Memes and Memes Beget Genes: Modeling a New Catalytic Closure," *Complexity* 4, no. 5 (1999): 22–28.
14. James N. Gardner, "The Selfish Biocosm: Complexity as Cosmology,"" *Complexity* 5, no. 3 (2000): 34–45.
15. Steven Weinberg, *The First Three Minutes* (New York: Basic Books, 1977), 148.
16. David Brewster, *Memoirs of the Life, Writing and Discoveries of Sir Isaac Newton*, vol. 2, ch. 27 (London: 1855).

Notes

Chapter One

1. Michael D. Lemonick, *Other Worlds: The Search for Life in the Universe* (New York: Simon & Schuster, 1998), 45.
2. Ian Crawford, "Where Are They?"" *Scientific American* (July 2000), 38–39.
3. Carl Sagan, *Contact* (New York: Pocket Books, 1986), 430–431.
4. Heinz R. Pagels, *The Dreams of Reason* (New York: Bantam, 1989).
5. Ibid., 155.
6. Ibid.
7. Ibid.
8. Ibid.
9. Ibid.
10. Ibid.
11. Ibid., 155–156.
12. Ibid., 156.
13. Ibid., 157.
14. Ibid.
15. Albert Einstein, http://www.aip.org/history/einstein/quantum2.htm (8 Dec. 2002).
16. Albert Einstein, quoted in Coey S. Powell, *God in the Equation* (New York: Free Press, 2002), 3.
17. Ibid., 252.
18. Ibid., 252–253.
19. Albert Einstein, *Ideas and Opinions* (New York: Three Rivers Press, 1982), 40.
20. Heinz R. Pagels, *Perfect Symmetry* (New York: Bantam, 1986), 378–379.
21. Heinz R. Pagels, *The Cosmic Code* (New York: Bantam, 1983).
22. Michael Polanyi, *Personal Knowledge* (Chicago: University of Chicago Press, 1958), 404.
23. Paul Davies, *The Fifth Miracle* (New York: Simon & Schuster, 1999), 246.
24. NASA's description of its Origins program is at http://origins.jpl.nasa.gov (8 Dec. 2002).
25. Ibid.
26. John A. Wheeler, *At Home in the Universe* (Woodbury: AIP Press, 1996), 310.
27. Pierre Teilhard de Chardin, *The Phenomenon of Man* (New York: Harper & Row, 1975), 287–288.
28. John D. Barrow and Frank J. Tipler, *The Anthropic Cosmological Principle* (Oxford: Oxford University Press, 1988).
29. Ibid., 677.

Chapter Two

1. Charles Darwin, *The Descent of Man* (Norwalk: Heritage Press, 1972), 328.
2. John D. Barrow, *The Book of Nothing* (New York: Pantheon, 2001).
3. Ibid., 9.
4. Ibid.
5. Ibid., 10.
6. Ibid.
7. Ibid., 298.
8. Ibid., 262.
9. Albert Einstein, *Ideas and Opinions* (New York: Three Rivers Press, 1954), 40.
10. John A. Wheeler, foreword to *The Anthropic Cosmological Principle*, by John D. Barrow and Frank J. Tipler (Oxford: Oxford University Press, 1988), vii.
11. Lee Smolin, "How Were the Parameters of Nature Selected?" http://online.itp.ucsb.edu/online/bblunch/smolin/oh/01.html (8 Dec. 2002).
12. P. A. M. Dirac, as quoted in Barrow and Tipler, *Anthropic Cosmological Principle*, 232.
13. Stephen Hawking, "Quantum Cosmology," in *The Nature of Space and Time*, by Stephen Hawking and Roger Penrose (Princeton: Princeton University Press, 1996), 89–90.
14. Julian Barbour, *The End of Time: The Next Revolution in Physics* (Oxford: Oxford University Press, 2000).
15. Ibid., 44.
16. Barrow and Tipler, *Anthropic Cosmological Principle*, 38.
17. Quoted in ibid., 59. For a more recent translation, see Benedict de Spinoza, *The Ethics and Other Works*, ed. and trans. Edwin Curley (Princeton: Princeton University Press, 1994), 111–112.
18. Erasmus Darwin, *Zoonomia, or the Laws of Organic Life* (London: 1794), quoted in Barrow and Tipler, *Anthropic Cosmological Principle*, 72.
19. William Paley, *Natural Theology* (London: 1802), quoted in Richard Dawkins, *The Blind Watchmaker* (New York: Norton, 1987), 4.
20. Barrow and Tipler, *Anthropic Cosmological Principle*, 80.
21. Ibid., 80.
22. Ibid.
23. Ibid., 81.
24. Lawrence J. Henderson, *The Fitness of the Environment* (Cambridge: Harvard University Press, 1913).
25. For a contemporary discussion of water's unique physical and chemical properties, see Wes Marrin, *Universal Water* (Maui: Inner Ocean Publishing, 2002).

26. Lawrence J. Henderson, *The Order of Nature* (Cambridge: Harvard University Press, 1917), 192.
27. Henderson, *Fitness of the Environment*, 312.
28. For a delightful account of the contemporary rehabilitation of Einstein's discarded notion of antigravity, see Corey S. Powell, *God in the Equation* (New York: Free Press, 2002).
29. Martin Rees, *Just Six Numbers* (New York: Basic Books, 2000).
30. Ibid., x.
31. Ibid., 30–31.
32. Ibid., 49.
33. Ibid.
34. Ibid., 88.
35. Ibid.
36. Ibid., 88–89.
37. Ibid., 99.
38. Ibid., 115.
39. Ibid.
40. Ibid., 106.
41. Stephen Hawking, *The Universe in a Nutshell* (New York: Bantam Books, 2001), 54.
42. Ibid., 87.
43. Rees, *Just Six Numbers*, 136.
44. Ibid., 161.
45. Steven Weinberg, "Can Science Explain Everything? Anything?"" *The New York Review of Books*, 31 May 2001, 49–50.
46. Albert Einsten, "Letter to Max Born of December 4, 1926," quoted in *Quantum Mechanics, 1925–1927: Triumph of the Copenhagen Interpretation*, http://www.aip.org/history/heisenberg/p09_text.htm (8 Dec. 2002).
47. John A. Wheeler, *At Home in the Universe*, 310.
48. Ibid., 185.
49. Ibid., 186.
50. The official Karl Popper website is located at http://www.eeng.dcu.ie/~tkpw/ (8 Dec. 2002).
51. John A. Wheeler, *At Home in the Universe*, 185–186.
52. Ibid., 296.
53. Ibid., 300.
54. Ibid., 300–301.

55. Ibid., 305–307.
56. Ibid., 187.
57. Barrow and Tipler, *Anthropic Cosmological Principle*, 614.
58. Ibid., 675–677.
59. Pagels, *Perfect Symmetry*, 377–378.

Chapter Three

1. Stuart Kauffman, *Investigations* (Oxford: Oxford University Press, 2000), xi.
2. L. Dyson, M. Kleban, and L. Susskind, "Disturbing Implications of a Cosmological Constant," 1 Aug. 2002, xxx.lanl.gov (arXiv: hep-th/0208013 v1) (8 Dec. 2002).
3. John Horgan, *The End of Science* (New York: Broadway Books, 1997), 132.
4. Stuart A. Kauffman, *The Origins of Order* (Oxford: Oxford University Press, 1993), xiii.
5. Ibid., xvi.
6. Ibid., 644.
7. Richard Dawkins, "The Evolution of Evolvability," in *Artificial Life,* vol. 6 (Santa Fe: Santa Fe Institute, 1988).
8. Stuart Kauffman, *At Home in the Universe* (Oxford: Oxford University Press, 1995).
9. Kauffman, *Investigations.*
10. Kauffman, *At Home in the Universe*, 8.
11. Ibid., 19–20.
12. Ibid., 20.
13. Kauffman, *Investigations*, 243–244.
14. Ibid., 85.
15. Ibid., 49.
16. Charles Darwin, *The Origin of Species* (New York: Random House, 1993), 636–637.
17. Francis Crick, "On Protein Synthesis," Symp. Soc. Exp. Biol. 12, 1958, 138–167.
18. Richard Dawkins, *The Extended Phenotype* (Oxford: Oxford University Press, 1989), 289.

Chapter Four

1. Thomas Henry Huxley, *A Liberal Education, Lay Sermons, and Reviews* (London: 1868), http://www.bartleby.com/66/11/30011.html (8 Dec. 2002).
2. *Edge* interview with Paul Davies, http://www.edge.org/3rd_culture/davies/davies_p6.html (8 Dec. 2002).
3. "Purpose of the Templeton Prize for Progress Toward Research or Discoveries about Spiritual Realities," http://www.templetonprize.org/purpose.html (8 Dec.

2002).

4. Craig Ventner, as quoted in "The Century of Biology," *Business Week* (August 31, 1998), 86.
5. Christian de Duve, *Vital Dust: Life as a Cosmic Imperative* (New York: Basic Books, 1995), 291.
6. Jacques Monod, *Chance and Necessity*, trans. A. Wainhouse (New York: Knopf, 1971), 145–146.
7. Ibid., 180.
8. de Duve, *Vital Dust*, 300.
9. Ibid., xviii.
10. Ibid., 297.
11. Ibid., 293.
12. Freeman Dyson, *Infinite in All Directions* (New York: Harper, 1988), 47.
13. Lee Smolin, *The Life of the Cosmos* (Oxford: Oxford University Press, 1997).
14. Steven Weinberg, *The First Three Minutes* (New York: Basic Books, 1977), 154.
15. Lawrence M. Krauss, "Cosmological Antigravity," *Scientific American* (January 1999): 59.
16. Neil Turok, "Before Inflation," 9 Nov. 2000, http://xxx.lanl.gov/PS_cache/astro-ph/pdf/0011/0011195.pdf (8 Dec. 2002).
17. Ibid.
18. Steven Weinberg, "A Designer Universe?" *The New York Review of Books*, 21 October 1999, 47.
19. Ibid.
20. Charles Darwin, *The Origin of Species* (New York: Random House, 1993), 649.
21. Ibid., 54.
22. Ibid.
23. Ibid.
24. Ibid., 57.
25. Ibid., 622.
26. John H. Holland, *Emergence: From Chaos to Order* (Reading, MA: Addison Wesley, 1998), 2.
27. Ibid., 5.
28. Ibid., 7.
29. Ibid.
30. Ibid.
31. Ibid., 9.
32. Ibid., 2.

33. Murray Gell-Mann, "What Is Complexity?" *Complexity* 1, no. 1 (1995): 16–17.
34. Richard Dawkins, *Climbing Mount Improbable* (New York: Norton, 1997).
35. Freeman J. Dyson, "Science & Religion: No Ends in Sight," *The New York Review of Books*, 28 March 2002, 6.
36. Michael Polanyi, *Personal Knowledge* (Chicago: University of Chicago Press, 1958), 404.
37. Edward O. Wilson, *Consilience: The Unity of Knowledge* (New York: Knopf, 1998).

Chapter Five

1. Ernst Mayr, "Darwin's Influence on Modern Thought," *Scientific American* (July 2000): 82.
2. Ibid., 82.
3. Ibid.
4. Ibid., 83.
5. Ibid., 81.
6. James Gleick, *Chaos: Making a New Science* (New York: Viking Penguin, 1988), 201.
7. Richard Dawkins, *River Out of Eden* (New York: Basic Books, 1995), 103–105.
8. Ibid., 105.
9. Ibid., 132–133.
10. Richard Dawkins, *The Selfish Gene* (Oxford: Oxford University Press, 1976).
11. Richard Dawkins, *The Blind Watchmaker: Why the Evidence of Evolution Reveals a Universe Without Design* (New York: Norton, 1986).
12. Steven Pinker, *The Blank Slate* (New York: Viking, 2002), 54.
13. Ibid.
14. Ibid.
15. Stuart A. Kauffman, *The Origins of Order* (Oxford: Oxford University Press, 1993), 177.

Chapter Six

1. Personal e-mail communication from Lee Smolin, 21 December 1999.
2. Lee Smolin, *The Life of the Cosmos* (Oxford: Oxford Universoty Press, 1997), 299.
3. Ernst Mayr, *One Long Argument: Charles Darwin and the Genesis of Modern Evolutionary Thought* (Cambridge: Harvard University Press, 1991), 85.

4. Peter D. Ward and Donald Brownlee, *Rare Earth: Why Complex Life Is Uncommon in the Universe* (New York: Copernicus, 2000).
5. Ibid., 156.
6. SETI Institute, "Are We So Special?" http://www.seti.org/science/rare_earth.html (8 Dec. 2002).
7. Martin Rees, *Before the Beginning: Our Universe and Others* (Reading, MA: Addison Wesley, 1997), 251.
8. John von Neumann, "On the General and Logical Theory of Automata," in John Casti, *Paradigms Lost* (New York: Avon, 1990), 131–132.
9. Richard Dawkins, "Implications of Natural Selection and the Laws of Physics" (Lee Smolin/Richard Dawkins colloquy), http://www.edge.org/discourse/index/cgi?OPTION=VIEW&THREAD=richard-dawkins/,,,smolin (8 Dec. 2002).

Chapter Seven

1. Michael J. Behe, *Darwin's Black Box* (New York: Touchstone, 1998), 39.
2. Eugenie Scott, "Scott Replies to Dembski," Feb. 2, 2002, http://www.natcenscied.org/resources/articles/3598_scott_replies_to_ dembski_2_2_2001.asp (8 Dec. 2002).
3. William A. Dembski, *The Design Inference* (Cambridge: Cambridge University Press, 1998), 26.
4. Ibid., 30.
5. William Dembski, "Teaching Intelligent Design—What Happened When? A Response to Eugenie Scott," http://www.trinity.edu/mbrown/godsmonsters&scientists/Texts/DembskiOnScott.html, 11 (8 Dec. 2002).
6. William A. Dembski, introduction to *Mere Creation: Science, Faith and Intelligent Design*, ed. William A. Dembski (Downers Grove: InterVarsity Press, 1998), 14.
7. Ibid., 29.
8. Scott, "Scott Replies to Dembski," 2.
9. Ibid.
10. Dembski, "Teaching Intelligent Design," 3.
11. Richard Dawkins, *River Out of Eden* (New York: Basic Books 1995), 133.
12. Steven Weinberg, "A Designer Universe?" *The New York Review of Books*, 21 October 1999, 48.
13. Ibid.
14. Frederick C. Crews, "Saving Us from Darwin," reprinted in *The Best American Science and Nature Writing of 2002*, ed. Tim Folger (Boston: Houghton Mifflin 2002), 43.

15. Niles Eldredge, *The Triumph of Evolution and the Failure of Creationism* (New York: W. H. Freeman, 2000), 81–89.
16. Stephen Jay Gould, *Full House* (New York: Harmony Books, 1996), 182–185.
17. Brian Greene, *The Elegant Universe* (New York: Norton, 1999), 370.
18. Daniel C. Dennett, *Darwin's Dangerous Idea: Evolution and the Meanings of Life* (New York: Touchstone, 1996), 176–177.
19. Martin Rees, *Our Cosmic Habitat* (Princeton: Princeton University Press, 2001), 170.
20. Paul Davies, *The Fifth Miracle* (New York: Simon & Schuster, 1999), 246.

Chapter Eight

1. John Casti, *Complexification* (New York: HarperCollins, 1994), 212.
2. Stephen Jay Gould, *Ontogeny and Phylogeny* (Cambridge: Harvard University Press, 1977), 13.
3. Ibid., 13–14.
4. Ibid., 2.
5. Ibid., 3.
6. Ibid.
7. Gerald M. Edelman, *Bright Air, Brilliant Fire: On the Matter of the Mind* (New York: Basic Books 1992), 22.
8. Charles Darwin, *The Origin of Species* (New York: Random House, 1993), 648–649.

Chapter Nine

1. Gerald M. Edelman, *Bright Air, Brilliant Fire: On the Matter of the Mind* (New York: Basic Books 1992), 203–204.
2. John Baez, online commentary on *The Life of the Cosmos,* http://www.aleph.se/Trans/Global/Omega/smolin/txt (8 Dec. 2002).
3. Richard Dawkins, "Implications of Natural Selection and the Laws of Physics" (Lee Smolin/Richard Dawkins colloquy), http://www.edge.org/discourse/index/cgi?OPTION=VIEW&THREAD=richard-dawkins/,,,smolin, 3 (8 Dec. 2002).
4. Martin Rees, *Before the Beginning: Our Universe and Others* (Reading, MA: Addison Wesley, 1997), 250.
5. Francis Crick, *Life Itself: Its Origin and Nature* (New York: Simon & Schuster, 1981.
6. Frank J. Tipler, *The Physics of Immortality* (New York: Doubleday, 1994), 55.
7. Ibid., 54.
8. Ibid., 55.

9. Ray Kurzweil, *The Age of Spiritual Machines: When Computers Exceed Human Intelligence* (New York: Viking, 1999).
10. Edelman, *Bright Air, Brilliant Fire,* 188–196.
11. Kurzweil, *The Age of Spiritual Machines,* 3.
12. Ibid., 5.
13. Ibid., 258–260.
14. John A. Wheeler, *At Home in the Universe* (Woodbury: AIP Press, 1996), 308.
15. John D. Barrow and Frank J. Tipler, *The Anthropic Cosmological Principle* (Oxford: Oxford University Press, 1988), 674–675.
16. Freeman Dyson, *Infinite in All Directions* (New York: Harper Perennial Library, 1988), 99–100.
17. Ibid., 117.
18. Ibid., 118.
19. Ibid.
20. Ibid., 118–119.
21. Richard Dawkins, *River Out of Eden* (New York: Basic Books, 1995), 136–136.
22. Ibid., 136–137.
23. Ibid., 138–139.
24. Ibid., 151.
25. Ibid., 158.
26. Richard Dawkins, foreword to Susan Blackmore, *The Meme Machine* (Oxford: Oxford University Press, 1999), viii.
27. Ibid., ix.
28. Daniel C. Dennett, *Consciousness Explained* (Boston: Little, Brown & Co., 1991), 202.
29. Dawkins, *River Out of Eden,* 160.

Chapter Ten

1. Freeman Dyson, *Disturbing the Universe* (New York: Harper & Row, 1979), 250.
2. Charles Darwin, "Letter to Asa Gray," 1860, http://www.pbs.org/wgbh/evolution/darwin/diary/1858.html (8 Dec. 2002).
3. Charles Darwin, *The Origin of Species* (New York: Random House, 1993), 638.
4. Frank J. Tipler, *The Physics of Immortality* (New York: Doubleday, 1994), 2.
5. George Johnson, *Fire in the Mind* (New York: Vintage, 1996), 221 (discussing thinking of Harold Morowitz).
6. Christian de Duve, *Life Evolving: Molecules, Mind, and Meaning* (Oxford: Oxford University Press, 2002), 288.

7. Stephen Jay Gould, *I Have Landed* (New York: Harmony Books, 2002), 217.
8. Ibid.
9. Daniel C. Dennett, *Consciousness Explained* (Boston: Little, Brown & Co., 1991), 202.
10. Michael J. Denton, *Nature's Destiny: How the Laws of Biology Reveal Purpose in the Universe* (New York: Free Press, 1998).
11. Ibid., xix.
12. Ibid., 386.
13. Alan H. Guth, *The Inflationary Universe: The Quest for a New Theory of Cosmic Origins* (Cambridge: Perseus Books, 1997), 253.
14. Ibid., 268.
15. Louis Crane, "On the Role of Intelligent Life in the Evolution of the Universe" (unpublished paper, Kansas State University, [undated]).
16. Martin Rees, *Before the Beginning: Our Universe and Others* (Reading, MA: Addison Wesley, 1997), 4.
17. Lee Smolin, *Three Roads to Quantum Gravity* (New York: Basic Books, 2001), vii.
18. Louis Crane, "Possible Implications of the Quantum Theory of Gravity," 17 Feb. 1994, http://xxx.lanl.gov/PS_cache/hep-th/pdf/9402/9402104.pdf (8 Dec. 2002).

Chapter Eleven

1. Gerald M. Edelman, *Bright Air, Brilliant Fire: On the Matter of the Mind* (New York: Basic Books, 1992).
2. Roger Penrose, *The Emperor's New Mind* (Oxford: Oxford University Press, 1989).
3. James N. Gardner, "Assessing the Robustness of the Emergence of Intelligence: Testing the Selfish Biocosm Hypothesis (#IAA-00-IAA.9.2.06)," *Acta Astronautica* 48, no. 5–12 (2001): 951–955.

Chapter Twelve

1. Freeman Dyson, "Time Without End: Physics and Biology in an Open Universe," *Review of Modern Physics* 51 (1979): 447–460.
2. Lawrence M. Krauss and Glenn D. Starkman, "Life, The Universe and Nothing: Life and Death in an Ever-Expanding Universe," 12 Feb. 1999, http://xxx.lanl.gov/PS_cache/astro-ph/pdf/9902/9902189.pdf (8 Dec. 2002).
3. Ibid.
4. Ibid.
5. Ibid., 2.
6. Ibid., 21.

7. Ibid.
8. Ibid.
9. Katherine Freese and William H. Kinney, "The Ultimate Fate of Life in an Accelerating Universe," 17 May 2002, http://xxx.lanl.gov/PS_cache/astro-ph/pdf/0205/0205279.pdf, 7 (8 Dec. 2002).
10. Justin Khoury, Burt A. Ovrut, Paul J. Steinhardt, and Neil Turok, "The Ekpyrotic Universe: Colliding Branes and the Origin of the Hot Big Bang," 29 Mar. 2001, http://xxx.lanl.gov/PS_cache/hep-th/pdf/0103/0103239.pdf, 2 (8 Dec. 2002).
11. Ibid., 4.
12. Ibid.
13. Paul J. Steinhardt and Neil Turok, "Cosmic Evolution in a Cyclic Universe," 12 Nov. 2001, http://xxx.lanl.gov/PS_cache/hep-th/pdf/0111/0111098.pdf (8 Dec. 2002).
14. Ibid.
15. Ibid.
16. Hong Liu, Gregory Moore, and Nathan Seiberg, "The Challenging Cosmic Singularity," 31 Dec. 2002, http://xxx.lanl.gov/PS_cache/gr-qc/pdf/0301/0301001.pdf (1 Feb. 2003).
17. Ibid.
18. Ibid.
19. Andrei Linde, "Inflationary Theory versus Ekpyrotic/Cyclic Scenario," 26 May 2002, http://xxx.lanl.gov/PS_cache/hep-th/pdf/0205/0205259.pdf (8 Dec. 2002).
20. Steinhardt and Turok, "Cosmic Evolution in a Cyclic Universe."
21. Seth Lloyd, "Ultimate Physical Limits to Computation," *Nature* 406 (2000): 1047–1054.
22. James N. Gardner, "Assessing the Computational Potential of the Eschaton: Testing the Selfish Biocosm Hypothesis," *Journal of the British Interplanetary Society* 55, no. 7/8 (2002), 285–288.
23. Frank J. Tipler, *The Physics of Immortality* (New York: Doubleday, 1994), 136.
24. Freeman Dyson, "Progress in Religion," speech of May 16, 2000, http://www.edge.org/3rd_culture/dyson_progress/dyson_progress_p2.html (8 Dec. 2002).
25. Michael Shermer, "Shermer's Last Law," *Scientific American* (January 2002): 33.

Chapter Thirteen

1. Andrei Linde, "Inflationary Theory versus Ekpyrotic/Cyclic Scenario," 26 May 2002, http://xxx.lanl.gov/PS_cache/hep-th/pdf/0205/0205259.pdf (8 Dec. 2002).

2. Fred Hoyle, "The Universe: Past and Present Reflections," *Annual Reviews of Astronomy and Astrophysics* 20 (1982), 16.
3. Fred Hoyle, *The Intelligent Universe: A New View of Creation and Evolution* (New York: Holt, Rinehart and Winston, 1984), 226.
4. Ibid., 237.
5. Ibid., 239.
6. Edward R. Harrison, "The Natural Selection of Universes Containing Intelligent Life," *Quarterly Journal of the Royal Astronomical Society* 36 (1995): 193–203.
7. Louis Crane, "On the Role of Intelligent Life in the Evolution of the Universe" (unpublished paper, Kansas State University, [undated]).
8. Ibid.
9. Charles Darwin, *Autobiography (1809–1882)* (New York: Norton, 1969), 92.
10. Francis Crick, *Life Itself: Its Origin and Nature* (New York: Simon & Schuster, 1981).
11. John Archibald Wheeler, *Geons, Black Holes and Quantum Foam* (New York: Norton, 1998), 355.
12. Harrison, "The Natural Selection of Universes Containing Intelligent Life," 199–200.
13. Paul J. Steinhardt and Neil Turok, "Cosmic Evolution in a Cyclic Universe," 12 Nov. 2001, http://xxx.lanl.gov/PS_cache/hep-th/pdf/0111/0111098.pdf (8 Dec. 2002).
14. J. Richard Gott III and Li-Xin Li, "Can the Universe Create Itself?" 30 Dec. 1997, http://xxx.lanl.gov/PS_cache/astro-ph/pdf/9712/9712344.pdf (8 Dec. 2002).
15. Ibid., 1.
16. Ibid., 2.
17. Ibid.
18. Ibid., 38.
19. Charles Darwin, *Autobiography,* 93.
20. John A. Wheeler, *At Home in the Universe* (Woodbury: AIP Press, 1996), 305–307.
21. Wheeler, *Geons, Black Holes and Quantum Foam,* 338.
22. Ibid., 337.
23. Ibid., 338.
24. *Edge* interview with Paul Davies, http://www.edge.org/3rd_culture/davies/davies_p6.html (8 Dec. 2002).

Chapter Fourteen

1. Stephen Hawking, "Quantum Cosmology, M-Theory and the Anthropic Principle," http://www.damtp.cam.ac.uk/hawking/quantum.html (8 Dec. 2002).
2. Brian Greene, *The Elegant Universe* (New York: Norton, 1999), 362.
3. Andrei Linde, "The Hard Art of Universe Creation," 14 Oct. 1991, http://xxx.lanl.gov/PS_cache/hep-th/pdf/9110/9110037.pdf (8 Dec. 2002).
4. Ibid.
5. Stephen Wolfram, *A New Kind of Science* (Champaign, IL: Wolfram Media, 2002).
6. Ibid., 465.
7. Ibid., 465–466.
8. Ibid., 471.
9. Peter D. Ward and Donald Brownlee, *Rare Earth: Why Complex Life Is Uncommon in the Universe* (New York: Copernicus, 2000), xv.

Chapter Fifteen

1. Martin Rees, *Just Six Numbers* (New York: Basic Books, 2000), 100.
2. Erwin Schrödinger, *What Is Life?* (Cambridge: Cambridge University Press, 1992), 22.

Chapter Sixteen

1. Stephen Jay Gould, *Full House* (New York: Three Rivers Press, 1997), 149–151.
2. Stephen Jay Gould, "The Pattern of Life's History," http://www.edge.org/3rd_culture/gould/gould_index.html (8 Dec. 2002).
3. For a lucid and entertaining account of the punctuated equilibrium controversy, see Niles Eldredge, *Reinventing Darwin: The Great Debate at the High Table of Evolutionary Theory* (New York: Wiley, 1995).
4. One critic who does challenge the prevailing orthodoxy regarding the absence of evidence of progress from the evolutionary record is British scientist Simon Conway Morris, whose essay "We Were Meant to Be" appeared in *New Scientist* 6, 16 November 2002, 26–29.
5. Gould, *Full House*, 4.
6. Ibid.
7. Ibid., 18.
8. Ibid., 29.
9. Ibid., 28.
10. Christian de Duve, *Life Evolving: Molecules, Mind, and Meaning* (Oxford: Oxford University Press, 2002), 297.
11. Robert Wright, *Nonzero: The Logic of Human Destiny* (New York: Pantheon, 1999), ix.

12. Ibid., xiii.
13. Gregory Stock, *Metaman: The Merging of Humans and Machines into a Global Superorganism* (New York: Simon & Schuster, 1993), 14.
14. Mark S. Frankel and Audrey R. Chapman, *Human Inheritable Genetic Modifications: Assessing Scientific, Ethical, Religious, and Policy Issues* (Washington, D.C.: American Association for the Advancement of Science, 2000).
15. James N. Gardner, "Genes Beget Memes and Memes Beget Genes: Modeling a New Catalytic Closure," *Complexity* 4, no. 5 (1999): 22–28.
16. Pierre Teilhard de Chardin, *The Phenomenon of Man* (New York: Harper Torchbooks, 1961).
17. Peter D. Medawar, "Critical Review of *The Phenomenon of Man*," *Mind* 70 (1961): 99–106.
18. Jacques Monod, *Chance and Necessity*, trans. A. Wainhouse (New York: Knopf, 1971), 32.
19. Harold J. Morowitz, *The Emergence of Everything* (Oxford: Oxford University Press, 2002), 15.
20. David Bohm and Basil J. Hiley, *The Undivided Universe* (London: Routledge, 1995).
21. John D. Barrow and Frank J. Tipler, *The Anthropic Cosmological Principle* (Oxford: Oxford University Press, 1988), 200–201.
22. The phenomenon is discussed at length in appendix 1.
23. James N. Gardner, "Genes Beget Memes and Memes Beget Genes."
24. Charles J. Lumsden and Edward O. Wilson, *Promethean Fire: Reflections on the Origin of Mind* (Cambridge: Harvard University Press, 1983), 170–171.
25. Chikara Furusawa and Kunihiko Kaneko, "Complex Organization in Multicellularity as a Necessity in Evolution," in *Proceedings of the Seventh International Conference on Artificial Life*, ed. Mark A. Bedau, John S. McCaskill, Norman H. Packard, and Steen Rasmussion (Cambridge: MIT Press, 2000), 111.
26. Stephen Wolfram, *A New Kind of Science* (Champaign, IL: Wolfram Media, 2002), 14.
27. Ibid., 388.
28. Ibid., 398–399.
29. Paul Davies, *The Fifth Miracle* (New York: Simon & Schuster, 1999), 259.
30. James E. Lovelock, *Gaia: A New Look at Life on Earth* (Oxford: Oxford University Press, 1989).
31. James E. Lovelock and Lynn Margulis, "Atmospheric Homeostasis By and For the Biosphere," *Tellus* 26 (1974): 2–10.
32. Keith Downing, "Exploring Gaia Theory: Artificial Life on a Planetary Scale," in *Proceedings of the Seventh International Conference on Artificial Life*, ed. Mark A.

Bedau, John S. McCaskill, Norman H. Packard, and Steen Rasmussion (Cambridge: MIT Press, 2000), 90.

33. Ibid.
34. Ibid.
35. Davies, *The Fifth Miracle*, 264–270.
36. Stephen Jay Gould, *Life's Grandeur* (London: Jonathan Cape, 1996), 167.
37. Ibid., 216.
38. Charles Darwin, *The Origin of Species* (New York: Random House, 1993), 648–649.

Chapter Seventeen

1. George B. Dyson, *Darwin Among the Machines: The Evolution of Global Intelligence* (Reading, MA: Helix Books, 1998), 175.
2. John R. Koza, Martin A. Keane, and Matthew J. Streeter, "Evolving Inventions," *Scientific American* (February 2003): 52–59.
3. Lewis Thomas, "Computers," *New England Journal of Medicine* 288, no. 24 (14 June 1973): 1289.
4. Nathaniel Hawthorne, *The House of the Seven Gables* (Columbus: Ohio State University Press, 1965), 264.
5. Lewis Thomas, "Social Talk," *New England Journal of Medicine* 287, no. 19 (9 November 1973): 974.
6. Freeman J. Dyson, "Science & Religion: No Ends in Sight," *The New York Review of Books*, 28 March 2002, 6.
7. Julian Huxley, "The New Divinity," http://www.xs4all.nl/~dap/Pagina's/Huxley_Julian.htm (8 Dec. 2002).

Chapter Eighteen

1. Stephen W. Hawking, *A Brief History of Time: From the Big Bang to Black Holes* (New York: Bantam, 1988), 175.
2. Paul Davies, *The Mind of God: The Scientific Basis for a Rational World* (New York: Touchstone, 1993), 232.
3. Ibid.
4. Immanuel Kant, "The Contest of Faculties" (1798), in *Political Writings*, ed. H. Reiss and trans. H. B. Nisbet (Cambridge: Cambridge University Press, 1970), 177.
5. Ibid.
6. Ibid., 183.
7. Ibid., 185.

8. Immanuel Kant, "Idea for a Universal History with a Cosmopolitan Purpose" (1784), in *Political Writings*, ed. H. Reiss and trans. H. B. Nisbet (Cambridge: Cambridge University Press, 1970), 41.
9. Ibid., 43.
10. Ibid., 44.
11. John Rawls, *A Theory of Justice* (Cambridge: Harvard University Press, 1971), 291.
12. Ibid., 284.
13. Ibid., 285.
14. Ibid., 290.
15. Ibid., 128.
16. Ibid., 284–293.
17. Charles J. Lumsden and Edward O. Wilson, *Promethean Fire: Reflections on the Origin of Mind* (Cambridge: Harvard University Press, 1983), 30.
18. Michael Shermer, "Why ET Hasn't Called," *Scientific American* (August 2002): 33.
19. Stephen Jay Gould, *Rocks of Ages: Science and Religion in the Fullness of Life* (New York: Ballantine, 1999).
20. Alfred North Whitehead, *Science and the Modern World* (New York: Free Press, 1967), 13.
21. Ibid., 12.
22. Ibid.
23. Ibid.
24. Menas Kafatos and Robert Nadeau, *The Conscious Universe: Parts and Wholes in Physical Reality* (New York: Springer, 2000), 145.
25. John A. Wheeler, foreword to *The Anthropic Cosmological Principle*, by John D. Barrow and Frank J. Tipler (Oxford: Oxford University Press, 1988), ix.
26. Michael Polanyi, *Personal Knowledge* (Chicago: University of Chicago Press, 1958), 405.
27. Christian de Duve, *Vital Dust: Life as a Cosmic Imperative* (New York: Basic Books, 1995), 301.

Afterword

1. Bertrand Russell, *Autobiography*, vol. 2, (London: George Allen and Unwin, 1968), 159.
2. Jacques Monod, *Chance and Necessity*, trans. A. Wainhouse (New York: Knopf, 1971), 180.

Glossary

Anthropic cosmological principle (or anthropic principle): The notion that the universe is fine-tuned in such a way as to be life-friendly. The principle has four principal versions: the *weak anthropic principle*, the *strong anthropic principle*, the *participatory anthropic principle*, and the *final anthropic principle*.

Anthropocentric cosmological viewpoint: The idea that the appearance of the human species is the fundamental end and purpose of the entire process of cosmic evolution.

Artificial life: A new science based on the hypothesis that the biochemical processes that energize living matter are actually elaborate forms of computation that are susceptible to simulation in computers.

Artificial selection: The key metaphor used by Charles Darwin to explain the appearance of a succession of genetically related species in the geological record by means of a process of natural selection. Artificial selection, in Darwin's view, encompassed two distinct practices: deliberate selective breeding of animal and plant species to achieve desired characteristics, and a form of selection, which he called unconscious, that results from animal breeders attempting to possess and breed from the best animals.

Astrobiology: The study of life in the universe, both on planet Earth and beyond.

Attractors: In complexity theory, a set of points or states in state space to which trajectories within some volume of state space converge asymptotically over time.

Baby universes: Hypothetical new universes that emerge from an existing universe under a wide range of hypothesized circumstances, including *eternal chaotic inflation*, *cosmological natural selection*, or the deliberate creation of a *universe in a lab* by a sufficiently advanced civilization.

Big Bang: The hypothesized beginning of our universe from a *singularity* in which matter and energy was compressed into a tiny space smaller than an atom and then exploded outward.

Big Bounce: The hypothesized transition of a collapsed universe from a *Big Crunch*, resulting in a new *Big Bang*.

Big Crunch: The hypothesized end of the universe in a massive collapse of the entire universe into a *singularity* that would bring an end to all of space and time.

Black hole: A place where the gravitational force is so intense that nothing, not even light, can escape. Astrophysicist Lee Smolin theorizes that new *baby universes* may emerge from the hearts of black holes.

Block universe: A theory of the nature of time and the universe in which every instant—past, present, and future—coexists simultaneously. Theorist Julian Barbour calls this mathematically perfect and eternal state *Platonia*.

Branes: A generic term for any of the extended objects postulated by *string theory* or *M-theory*, including 1-branes (one-dimensional objects), 2-branes (two-dimensional objects), etc.

Calabi-Yau shapes: Calabi-Yau spaces are the six-dimensional geometrical shapes hypothesized by *M-theory* in which compactified spatial dimensions beyond the three familiar extended spatial dimensions are curled up tightly.

Categorical imperative: A principle of ethics formulated by Immanuel Kant that one should act only on that maxim by which one can at the same time will that it should become a universal law.

Chronology protection conjecture: Conjecture by cosmologist Stephen Hawking that the laws of physics forbid the manifestation of *closed timelike curves*, at least at the macroscopic scale.

Closed timelike curves: Hypothetical configurations of space and time permitted under Albert Einstein's theory of relativity where gravity is sufficiently strong to bend the space-time continuum into a looping configuration that allows future events to influence the past.

Closed universe scenario: A hypothetical scenario in which the expansion of the universe will eventually slow down, halt, and then reverse as the universe begins to contract toward a *Big Crunch*.

Coevolution: The simultaneous and parallel evolution of two or more categories of *replicators* (including but not limited to biological organisms) where variations and adaptations in one set of coevolving replicators will evoke an adaptive response in the other set or sets of coevolving replicators.

Complexification: The progression of a system or entity over time from a state of lower internal diversity and complexity to a state of higher internal diversity and complexity.

Complexologists: The generic term for scientists like Stuart Kauffman and John Casti who study the behavior of complex adaptive systems.

Conscious artifact: An artificially created entity (such as a sufficiently advanced computer) that has acquired consciousness or sentience.

Consilience: The holistic notion associated with Harvard biologist Edward O. Wilson that all branches of human knowledge are fundamentally linked and constitute a deeply coherent, self-reinforcing intellectual system.

Copenhagen interpretation: One of the queerest aspects of quantum theory is that tiny subatomic particles like electrons are presumed to exist in a ghostlike state called a superposition until they are actually observed. It is only upon observation or measurement that such particles are forced to "choose" their key characteristics. Under the Copenhagen interpretation of quantum theory (named for the birthplace of quantum physics), the very act of measurement and observation affects basic aspects of subatomic reality.

Cosmic code: The term used by the late physicist Heinz Pagels to refer to the full suite of fundamental physical laws and constants that apply in our universe.

Cosmogenesis: The birth of a universe.

Cosmological constant: An idea put forward by Albert Einstein that the universe possesses a mysterious form of energy that functions like a kind of antigravity. Although Einstein later disavowed this idea, it has recently been resurrected with new terminology like *quintessence* and *lambda* and *dark energy.*

Cosmological natural selection: The concept pioneered by astrophysicist Lee Smolin that a Darwinian principle applies to the process of creation of *baby universes,* whereby those baby universes most adept at creating *black holes* will produce the most "offspring" in the form of new baby universes.

Curled-up dimensions: Dimensions higher than the four familiar *extended dimensions* (including time) that are too small to be detected with currently available scientific instruments.

Dark energy: A hypothesized antigravitational force that causes the expansion of the universe to speed up. Also known as *quintessence* or *lambda.*

Directed panspermia: The conjecture by Nobel Prize winner and DNA codiscoverer Francis Crick that intelligent aliens deliberately seeded life on Earth many billions of years ago.

DNA: The molecular device used by all life-forms on Earth to accomplish the feat of what Charles Darwin called "inheritance."

Drake equation: In 1961, Dr. Frank Drake, a pioneering researcher in the field of *SETI* (the Search for Extraterrestrial Intelligence), developed a formula that specifies the various factors that make the existence of technological civilizations beyond Earth more or less probable.

Efficient cause: A term in Aristotelian philosophy that corresponds roughly to contemporary notions of causation.

Ekpyrotic universe scenario: A new cosmic theory so named because of its resemblance to a particular vision of the cosmos articulated by the ancient Greeks. As described by its proponents, the term *ekpyrotic* is drawn from the Stoic model of cosmic evolution in which the universe is consumed by fire at regular intervals and reconstituted out of this fire, a conflagration called "ekpyrosis." The scenario proposes that the universe as we know it is made (and, perhaps, has been remade) through a conflagration ignited by collisions between *branes* along a hidden fifth dimension.

Emergence: The quality of phenomena or systems in which the behavior of the whole is much more complex than the behavior of the parts.

Entelechy: A force postulated by Aristotle that supposedly attracts the evolution of the cosmos toward a final state of perfection.

Entropy: The measure of the randomness of a particular system.

Epigenesis: A process by which morphological complexity develops gradually during embryology.

Eschaton: A hypothesized final cosmic state that would be capable of functioning as a duplicating device in the context of the *Selfish Biocosm hypothesis.*

Eternal chaotic inflation: Cosmologist Andrei Linde's theory that the Big Bang is not a one-time affair but rather a single instance of an eternal process of new-universe creation.

Eukaryotic: Organisms consisting of large cells having a separated nucleus and cytoplasm containing organelles like mitochondria.

Evolution: The change in a system over time, often including features such as *emergence* and *natural selection.*

Extended dimensions: The four familiar dimensions (including time) that are large and perceptible, in contrast to the tiny *curled-up dimensions* postulated by *string theory.*

Falsifiability: Scientific shorthand for the empirical testability of new hypotheses and their implications. Falsifiability is the hallmark of genuine science, sharply demarcating it from other arenas of human thought and experience like religion, mysticism, and metaphysics.

Fermi paradox: Named after the physicist Enrico Fermi, who, noting the conspicuous absence of any indication of the existence of intelligent alien life, commented during a luncheon conversation in Los Alamos in 1950, "Where are they?" The evident absence of evidence of extraterrestrial life was paradoxical, in Fermi's view, if the laws of nature were presumed to be inherently life-friendly.

Final anthropic principle: The final anthropic principle advances the extraordinary claim that once life has arisen anywhere in this or any other universe, its sophistication and pervasiveness will expand inexorably and exponentially until life's domain is coterminous with the boundaries of the cosmos itself.

Final cause: The teleological notion in Aristotelian philosophy that every entity or process is drawn irresistibly toward the purpose of that object.

Fourth Law of Thermodynamics: A possible new law of thermodynamics for self-constructing open thermodynamic systems such as biospheres that has been hypothesized by complexity theorist Stuart Kauffman, who thinks that this possible new law may capture a fundamental and hitherto unrecognized force in the universe that relentlessly drives the cosmos as a whole to construct itself to be as complex and diverse as possible. This suspected fourth law would serve as a counterforce to the disorder-generating *Second Law of Thermodynamics.*

Gaia hypothesis: The theory proposed by British atmospheric scientist James E. Lovelock and refined with the assistance of biologist Lynn Margulis that both animate and inanimate elements of Earth's environment are linked in a series of feedback loops.

Genetic algorithms: Advanced forms of software that employ the Darwinian tool of natural selection to evolve ever greater computational skills.

Heat death of the universe: The hypothetical final state of cosmic evolution that would occur in the far distant future when all thermodynamic processes have exhausted themselves and the cosmos settles into a bland, featureless uniformity.

Human germline therapy: The inheritable alteration of the human genetic sequence in reproductive cells to either avoid a predicted disability in progeny or enhance a particular trait associated with a certain gene.

Inflation: The prevailing cosmological theory that our universe underwent an episode of *superluminal* expansion immediately after the *Big Bang.*

Intelligent design: The idea that a system or phenomena reveals evidence of having been designed by an *intelligent designer.* Critics of the concept (including most adherents of the theory of Darwinian evolution through natural selection) consider it a thinly veiled version of creationism.

Intelligent designer: The entity or entities purportedly responsible for the *intelligent design* of nature and natural systems or processes.

Inverse square law of gravitation: The basic law of gravity, first fully articulated by Isaac Newton, which provides that the force of gravity diminishes in accordance with the square of the distance separating two gravitating bodies.

Irreducible complexity: The quality exhibited by a single system composed of several well-matched, interacting parts that contribute to its basic functioning, wherein the removal of any one of the parts causes the system to effectively cease functioning.

Lambda: A hypothesized antigravitational force that causes the expansion of the universe to speed up. Also known as *quintessence* or *dark energy.*

Mediocrity, principle of: The principle of mediocrity is a statistically based rule of thumb that, absent contrary evidence, a particular sample (Earth, for instance, or our universe) should be assumed to be a typical example of the ensemble of which it is a part.

Meliorism: The idea that biological evolution exhibits gradual improvement while progressing toward a final state of perfection intended by an intelligent designer.

Memes: Hypothetical units of cultural transmission. The term was coined by British evolutionary theorist Richard Dawkins to designate the cultural counterparts of genes, which are the basic units of genetic transmission.

Metaman: A hypothetical superorganism encompassing all of human civilization on Earth as well as the artifacts of that civilization.

Mitochondria: Endosymbiont-derived organelles present in most eukaryotic cells that are responsible for energy and that are believed to have descended from free-living bacteria.

Modified ekpyrotic cyclic universe scenario: In this modified version of the *ekpyrotic universe scenario*, the universe is hypothesized to undergo an endless sequence of cosmic epochs that begin with the universe expanding from a *Big Bang* and end with the universe contracting to a *Big Crunch*.

M-theory: A refinement of *string theory* that encompasses theoretical structures other than strings, including *branes*.

Multiverse: The notion associated with Andrei Linde's theory of *eternal chaotic inflation* that our universe is only one member of a vast ensemble of disconnected universes, each of which may have different laws of physics.

Neural Darwinism: The hypothesis of Nobel laureate Gerald Edelman that the development of the brain is a selectional process, similar to Darwinian selection, involving populations of neurons engaged in life-and-death competition in the maturing human embryo.

Nonergodic: A nonuniform, nonequilibrium state.

Omega Point: The concept of Omega Point (or Point Omega) is associated with the religious philosopher Pierre Teilhard de Chardin and, more recently, with the work of cosmologists John Barrow and Frank Tipler. For Teilhard, the Omega Point was the predicted end point of terrestrial evolution—an epoch in the distant future when humanity will have evolved into a planetwide superorganism that would coincide with the incarnation of the Christian God in the physical universe. Teilhard's overtly religious conception of the Omega Point is distinct from the secular version articulated by Barrow and Tipler in their book *The Anthropic Cosmological Principle*. For them, the Omega Point is the hypothesized final point in the evolution of a linked set of closed universes (a so-called *multiverse*) that proceed to contract toward a Big Crunch billions of years hence. Assuming that life persists and continues to evolve until the Omega Point, Barrow and Tipler project that life's counterentropic, or order-generating, organizational power will then come to dominate inanimate matter and energy completely.

Ontogeny: The developmental history of an individual organism.

Ontogon: A term coined by the author to designate a hypothetical eternal wave function responsible for the process of biological information generation that moves along a closed timelike curve from the past to the future and back again across the Big Crunch era to a new Big Bang era without disruption by a final singularity.

Open universe scenario: A hypothetical scenario in which the universe will expand indefinitely and at an ever increasing rate, yielding an essentially empty cosmos in the far distant future.

Participatory anthropic principle: The counterintuitive participatory anthropic principle hypothesizes, on the strength of the *Copenhagen interpretation* of quantum mechanics, that observer-participancy is necessary to summon the universe into existence and to give it structure.

Phylogeny: The evolutionary history or sequence of a lineage of organisms linked by descent from a common ancestor.

Platonia: A term coined by theorist Julian Barbour to refer to the timeless landscape postulated in his *block universe* theory where nothing changes and where all the illusory instants of time coexist simultaneously.

Punctuated equilibrium: The theory associated with Stephen Jay Gould and Niles Eldredge that evolution is characterized by long-term species stability, interrupted at rare intervals by episodes of rapid macroevolution.

Quantum mechanics: Fundamental theory of nature applicable at the scale of atoms and subatomic particles. This counterintuitive theory was developed at the beginning of the twentieth century by European scientists, including Niels Bohr, Albert Einstein, Werner Heisenberg, Erwin Schrödinger, and others.

Quintessence: A hypothesized antigravitational force that causes the expansion of the universe to speed up. Also known as *lambda* or *dark energy*.

Radial energy: A form of energy hypothesized by Teilhard de Chardin that drives the entire biosphere to ever higher levels of complexity and sophistication.

Relativity, general: Albert Einstein's theory that gravity is explicable as the curvature of the space-time continuum.

Replicators: Reproducing entities, including genes and *memes.*

Santa Fe Institute: The informal global headquarters for the scientific study of complex adaptive systems, located in Santa Fe, New Mexico.

Second Law of Thermodynamics: The Second Law of Thermodynamics states that *entropy* in a closed system can never decrease.

Selfish Biocosm hypothesis: A new theory first articulated by the author in a 2000 essay in the journal *Complexity* that proposes that the anthropic qualities that our universe exhibits can be explained as incidental consequences of an enormously lengthy cosmic replication cycle in which a cosmologically extended biosphere provides the means by which our cosmos duplicates itself and propagates one or more *baby universes.* The hypothesis suggests that the cosmos is "selfish" in the same metaphorical sense that evolutionary theorist Richard Dawkins proposed that genes are "selfish." Under the theory, the cosmos is "selfishly" focused upon the overarching objective of achieving its own replication. To use the terminology favored by economists, self-reproduction is the hypothesized *utility function* of the universe.

SETI Institute: The informal global headquarters for the scientific search for extraterrestrial intelligence located in Mountain View, CA.

SETI: An acronym for the search for extraterrestrial intelligence.

Singularity: A theoretical place at the center of a *black hole* where space is infinitely compressed and time is infinitely dilated.

Somatic gene therapy: Noninheritable genetic changes induced in somatic cells in order to treat genetically based diseases.

Stellar nucleosynthesis: The process by which heavy elements are forged from lighter elements in powerful thermonuclear reactions that take place in the interiors of giant stars and in the explosions of titanic supernovae.

String theory: A theory that the fundamental particles are actually different modes of vibration of tiny one-dimensional strings vibrating in eleven-dimensional space-time.

Strong anthropic principle: The controversial strong anthropic principle predicts that the origin of life and intelligence in the universe will eventually be shown to be strongly favored or even predestined by the laws and constants of inanimate nature.

Super-Copernican principle: Derived from the Copenhagen interpretation of quantum physics by physicist John Wheeler, this principle rejects the notion that the present (as opposed to the past or the future) is in any respect temporally special or privileged, just as Copernicus rejected the idea that Earth occupied a special or privileged position in the universe.

Superluminal: Faster than the speed of light.

Symbiogenesis: The notion that evolution proceeds in part through the appearance of symbiotic arrangements.

Teleology: The idea, associated primarily with the philosophy of Aristotle, that the evolutionary development of a system is drawn irresistibly toward a so-called *final cause.*

Teleonomy: The idea that the DNA molecules that comprise an organism's genome prescribe the organism's future ontogenetic pathway and the form of its mature phenotype.

Thermodynamics: Laws describing heat, energy, and *entropy* in physical systems.

Transgenerational moral imperative: The obligation to build a moral as well as a technological foundation for the benefit of the lives and minds yet to come. This concept is associated with the American moral philosopher John Rawls.

Ultimate computer: As described by computer scientist Seth Lloyd, a computing device as powerful as the laws of physics will allow.

Utility function: A term drawn from economics that specifies the outcome that is maximized by a particular process.

Weak anthropic principle: The less controversial weak version of the anthropic principle merely states in tautological fashion that since human observers inhabit this particular universe, it must perforce be life-friendly or it would not contain any observers resembling ourselves.

Bibliography

Barbour, Julian. *The End of Time: The Next Revolution in Physics.* Oxford: Oxford University Press, 2000.

Barrow, John D. *The Book of Nothing.* New York: Pantheon, 2001.

Barrow, John D., and Frank J. Tipler. *The Anthropic Cosmological Principle.* Oxford: Oxford University Press, 1988.

Behe, Michael J. *Darwin's Black Box.* New York: Touchstone, 1998.

Bohm, David, and Basil J. Hiley. *The Undivided Universe.* London: Routledge, 1995.

Brewster, David. *Memoirs of the Life, Writing and Discoveries of Sir Isaac Newton. Vol. 2,* ch. 27. London: 1855.

Casti, John. *Complexification.* New York: HarperCollins, 1994.

———. *Paradigms Lost.* New York: Avon, 1990.

———. *Would-Be Worlds.* New York: Wiley, 1997.

Crane, Louis. "Possible Implications of the Quantum Theory of Gravity." 17 Feb. 1994. http://xxx.lanl.gov/PS_cache/hep-th/pdf/9402/9402104.pdf (8 Dec. 2002).

Crawford, Ian. "Where Are They?" *Scientific American* (July 2000): 38–39.

Crews, Frederick C. "Saving Us from Darwin." Reprinted in *The Best American Science and Nature Writing of 2002,* edited by Tim Folger. Boston: Houghton Mifflin, 2002.

Crick, Francis. *Life Itself: Its Origin and Nature.* New York: Simon & Schuster, 1981.

Darwin, Charles. *Autobiography (1809–1882).* New York: Norton, 1969.

———. *The Descent of Man.* Norwalk: Heritage Press, 1972.

———. *The Origin of Species.* New York: Random House, 1993.

Darwin, Erasmus. *Zoonomia, or the Laws of Organic Life.* London: 1794.

Davies, Paul. *The Fifth Miracle.* New York: Simon & Schuster, 1999.

———. *The Mind of God: The Scientific Basis for a Rational World.* New York: Touchstone, 1993.

Dawkins, Richard. *The Blind Watchmaker: Why the Evidence of Evolution Reveals a Universe Without Design.* New York: Norton, 1986.

———. *Climbing Mount Improbable*. New York: Norton, 1997.

———. "The Evolution of Evolvability." In *Artificial Life*. Vol. 6. Santa Fe: Santa Fe Institute, 1988.

———. *River Out of Eden*. New York: Basic Books, 1995.

———. *The Selfish Gene*. Oxford: Oxford University Press, 1976.

De Duve, Christian. *Life Evolving: Molecules, Mind, and Meaning*. Oxford: Oxford University Press, 2002.

———. *Vital Dust: Life as a Cosmic Imperative*. New York: Basic Books, 1995.

Dembski, William A. *The Design Inference*. Cambridge: Cambridge University Press, 1998.

———. Introduction to *Mere Creation: Science, Faith and Intelligent Design*, edited by William A. Dembski. Downers Grove: InterVarsity Press, 1998.

Dennett, Daniel C. *Consciousness Explained*. Boston: Little, Brown & Co., 1991.

———. *Darwin's Dangerous Idea: Evolution and the Meanings of Life*. New York: Touchstone, 1996.

Denton, Michael J. *Nature's Destiny: How the Laws of Biology Reveal Purpose in the Universe*. New York: Free Press, 1998.

Downing, Keith. "Exploring Gaia Theory: Artificial Life on a Planetary Scale." In *Proceedings of the Seventh International Conference on Artificial Life*, edited by Mark A. Bedau, John S. McCaskill, Norman H. Packard, and Steen Rasmussion. Cambridge: MIT Press, 2000.

Dyson, Freeman. *Disturbing the Universe*. New York: Harper & Row, 1979.

———. *Infinite in All Directions*. New York: Harper Perennial Library, 1988.

———. "Science & Religion: No Ends in Sight." *The New York Review of Books*, 28 March 2002.

———. "Time Without End: Physics and Biology in an Open Universe." *Review of Modern Physics* 51 (1979): 447–460.

Dyson, George B. *Darwin Among the Machines: The Evolution of Global Intelligence*. Reading, MA: Helix Books, 1998.

Dyson, L., M. Kleban, and L. Susskind. "Disturbing Implications of a Cosmological Constant." 1 Aug. 2002. xxx.lanl.gov (arXiv: hep-th/0208013 v1) (8 Dec. 2002).

Edelman, Gerald M., *Bright Air, Brilliant Fire: On the Matter of the Mind*. New York: Basic Books, 1992.

Einstein, Albert. *Ideas and Opinions*. New York: Three Rivers Press, 1982.

Eldredge, Niles. *Reinventing Darwin: The Great Debate at the High Table of Evolutionary Theory*. New York: Wiley, 1995.

———. *The Triumph of Evolution and the Failure of Creationism*. New York: W. H. Freeman, 2000.

Freese, Katherine, and William H.Kinney. "The Ultimate Fate of Life in an Accelerating Universe." 17 May 2002. http://xxx.lanl.gov/PS_cache/astro-ph/pdf/0205/0205279.pdf (8 Dec. 2002).

Furusawa, Chikara, and Kunihiko Kaneko. "Complex Organization in Multicellularity as a Necessity in Evolution." In *Proceedings of the Seventh International Conference on Artificial Life,* edited by Mark A. Bedau, John S. McCaskill, Norman H. Packard, and Steen Rasmussion. Cambridge: MIT Press, 2000.

Gardner, James N. "Assessing the Computational Potential of the Eschaton: Testing the Selfish Biocosm Hypothesis." *Journal of the British Interplanetary Society* 55, no. 7/8 (2002): 285–288.

———. "Assessing the Robustness of the Emergence of Intelligence: Testing the Selfish Biocosm Hypothesis" (#IAA-00-IAA.9.2.06). *Acta Astronautica* 48, no. 5–12 (2001): 951–955.

———. "Genes Beget Memes and Memes Beget Genes: Modeling a New Catalytic Closure." *Complexity* 4, no. 5 (1999): 22–28.

———. "Mastering Chaos at History's Frontier: The Geopolitics of Complexity." *Complexity* 3, no. 2 (1997): 28–32.

———. "The Selfish Biocosm: Complexity as Cosmology." *Complexity* 5, no. 3 (2000): 34–45.

Gell-Mann, Murray. "What Is Complexity?" *Complexity* 1, no. 1 (1995): 16–17.

Gleick, James. *Chaos: Making a New Science.* New York: Viking Penguin, 1988.

Gold, Thomas. *The Deep Hot Biosphere.* New York: Copernicus, 2001.

Gott, J. Richard, III. *Time Travel in Einstein's Universe.* New York: Houghton Mifflin, 2001.

Gott, J. Richard, III, and Li-Xin Li. "Can the Universe Create Itself?" 30 Dec. 1997. http://xxx.lanl.gov/PS_cache/astro-ph/pdf/9712/9712344.pdf (8 Dec. 2002).

Gould, Stephen Jay. *Full House.* New York: Three Rivers Press, 1997.

———. *I Have Landed.* New York: Harmony Books, 2002.

———. *Life's Grandeur.* London: Jonathan Cape, 1996.

———. *Ontogeny and Phylogeny.* Cambridge: Harvard University Press, 1977.

———. *Rocks of Ages: Science and Religion in the Fullness of Life.* New York: Ballantine, 1999.

Greene, Brian. *The Elegant Universe.* New York: Norton, 1999.

Guth, Alan H. *The Inflationary Universe: The Quest for a New Theory of Cosmic Origins.* Cambridge: Perseus Books, 1997.

Harrison, Edward R. "The Natural Selection of Universes Containing Intelligent Life." *Quarterly Journal of the Royal Astronomical Society* 36 (1995): 193–203.

Hawking, Stephen W. *A Brief History of Time: From the Big Bang to Black Holes.* New York: Bantam, 1988.

———. *The Universe in a Nutshell.* New York: Bantam Books, 2001.

Hawking, Stephen, and Roger Penrose. *The Nature of Space and Time.* Princeton: Princeton University Press, 1996.

Hawthorne, Nathaniel. *The House of the Seven Gables.* Columbus: Ohio State University Press, 1965.

Henderson, Lawrence J. *The Fitness of the Environment.* Cambridge: Harvard University Press, 1913.

———. *The Order of Nature.* Cambridge: Harvard University Press, 1917.

Holland, John H. *Emergence: From Chaos to Order.* Reading, MA: Addison Wesley, 1998.

Horgan, John. *The End of Science.* New York: Broadway Books, 1997.

Hoyle, Fred. *The Intelligent Universe: A New View of Creation and Evolution.* New York: Holt, Rinehart and Winston, 1984.

———. "The Universe: Past and Present Reflections." *Annual Reviews of Astronomy and Astrophysics* 20 (1982).

Johnson, George. *Fire in the Mind.* New York: Vintage, 1996.

Kafatos, Menas, and Robert Nadeau. *The Conscious Universe: Parts and Wholes in Physical Reality.* New York: Springer, 2000.

Kant, Immanuel. "The Contest of Faculties" (1798). In *Political Writings,* edited by H. Reiss and translated by H. B. Nisbet. Cambridge: Cambridge University Press, 1970.

———. "Idea for a Universal History with a Cosmopolitan Purpose" (1784). In *Political Writings,* edited by H. Reiss and translated by H. B. Nisbet. Cambridge: Cambridge University Press, 1970.

Kauffman, Stuart. *At Home in the Universe.* Oxford: Oxford University Press, 1995.

———. *Investigations.* Oxford: Oxford University Press, 2000.

———. *The Origins of Order.* Oxford: Oxford University Press, 1993.

Khoury, Justin, Burt A. Ovrut, Paul J. Steinhardt, and Neil Turok. "The Ekpyrotic Universe: Colliding Branes and the Origin of the Hot Big Bang." 29 Mar. 2001. http://xxx.lanl.gov/PS_cache/hep-th/pdf/0103/0103239.pdf (8 Dec. 2002).

Krauss, Lawrence M. "Cosmological Antigravity." *Scientific American* (January 1999): 59.

Krauss, Lawrence M., and Glenn D. Starkman. "Life, the Universe and Nothing: Life and Death in an Ever-Expanding Universe." 12 Feb. 1999. http://xxx.lanl.gov/PS_cache/astro-ph/pdf/9902/9902189.pdf (8 Dec. 2002).

Kuhn, Thomas S. *The Structure of Scientific Revolutions.* Chicago: University of Chicago Press, 1970.

Kurzweil, Ray. *The Age of Spiritual Machines: When Computers Exceed Human Intelligence.* New York: Viking, 1999.

Linde, Andrei. "The Hard Art of Universe Creation." 14 Oct. 1991. http://xxx.lanl.gov/PS_cache/hep-th/pdf/9110/9110037.pdf (8 Dec. 2002).

———. "Inflationary Theory versus Ekpyrotic/Cyclic Scenario." 26 May 2002. http://xxx.lanl.gov/PS_cache/hep-th/pdf/0205/0205259.pdf (8 Dec. 2002).

Lloyd, Seth. "Ultimate Physical Limits to Computation." Nature 406 (2000): 1047–1054.

Lovelock, James E. Gaia: *A New Look at Life on Earth.* Oxford: Oxford University Press, 1989.

Lovelock, James E., and Lynn Margulis. "Atmospheric Homeostasis By and For the Biosphere." *Tellus* 26 (1974): 2–10.

Lumsden, Charles J., and Edward O. Wilson. *Promethean Fire: Reflections on the Origin of Mind.* Cambridge: Harvard University Press, 1983.

Marrin, Wes. *Universal Water.* Maui: Inner Ocean Publishing, 2002.

Mayr, Ernst. "Darwin's Influence on Modern Thought." *Scientific American* (July 2000).

Medawar, Peter D. "Critical Review of *The Phenomenon of Man.*" *Mind* 70 (1961): 99–106.

Miller, Kenneth R. *Finding Darwin's God.* New York: Cliff Street Books, 2000.

Monod, Jacques. *Chance and Necessity.* Translated by A. Wainhouse. New York: Knopf, 1971.

Morowitz, Harold J. *The Emergence of Everything.* Oxford: Oxford University Press, 2002.

Morris, Simon Conway. "We Were Meant to Be." *New Scientist* 6, 16 November 2002 26–29.

Pagels, Heinz R. *The Cosmic Code.* New York: Bantam, 1983.

———. *The Dreams of Reason.* New York: Bantam, 1989.

———. *Perfect Symmetry.* New York: Bantam, 1986.

Paley, William. *Natural Theology.* London: 1802.

Pinker, Steven. *The Blank Slate.* New York: Viking, 2002.

Polanyi, Michael. *Personal Knowledge.* Chicago: University of Chicago Press, 1958.

Powell, Corey S. *God in the Equation.* New York: Free Press, 2002.

Rawls, John. *A Theory of Justice.* Cambridge: Harvard University Press, 1971.

Rees, Martin. *Before the Beginning: Our Universe and Others.* Reading, MA: Addison Wesley, 1997.

———. *Just Six Numbers: The Deep Forces That Shape the Universe.* New York: Basic Books, 2000.

———. *Our Cosmic Habitat.* Princeton: Princeton University Press, 2001.

Russell, Bertrand. *Autobiography.* Vol. 2. London: George Allen and Unwin, 1968.

Sagan, Carl. *Contact.* New York: Pocket Books, 1986.

Schrödinger, Erwin. *What Is Life?* Cambridge: Cambridge University Press, 1992.

Shermer, Michael. *The Borderlands of Science.* Oxford: Oxford University Press, 2001.

———. "Shermer's Last Law." *Scientific American* (January 2002).

———. "Why ET Hasn't Called." *Scientific American* (August 2002).

Smolin, Lee. *The Life of the Cosmos.* Oxford: Oxford University Press, 1997.

———. *Three Roads to Quantum Gravity.* New York: Basic Books, 2001.

Spinoza, Benedict de. *The Ethics and Other Works.* Edited and translated by Edwin Curley. Princeton: Princeton University Press, 1994.

Steinhardt, Paul J., and Neil Turok. "Cosmic Evolution in a Cyclic Universe." 12 Nov. 2001. http://xxx.lanl.gov/PS_cache/hep-th/pdf/0111/0111098.pdf (8 Dec. 2002).

Stock, Gregory. *Metaman: The Merging of Humans and Machines into a Global Superorganism.* New York: Simon & Schuster, 1993.

Teilhard de Chardin, Pierre. *The Phenomenon of Man.* New York: Harper & Row, 1975.

Thomas, Lewis. "Computers." *New England Journal of Medicine* 288, no. 24 (14 June 1973).

———. "Social Talk." *New England Journal of Medicine* 287, no. 19 (9 November 1973).

Tipler, Frank J. *The Physics of Immortality.* New York: Doubleday, 1994.

Turok, Neil. "Before Inflation." 9 Nov. 2000. http://xxx.lanl.gov/PS_cache/astro-ph/pdf/0011/0011195.pdf (8 Dec. 2002).

Ward, Peter D., and Donald Brownlee. *Rare Earth: Why Complex Life Is Uncommon in the Universe.* New York: Copernicus, 2000.

Weinberg, Steven. "Can Science Explain Everything? Anything?" *The New York Review of Books,* 31 May 2001.

———. "A Designer Universe?" *The New York Review of Books,* 21 October 1999.

———. *The First Three Minutes.* New York: Basic Books, 1977.

Wheeler, John A. *At Home in the Universe.* Woodbury: AIP Press, 1996.

———. *Geons, Black Holes and Quantum Foam.* New York: Norton, 1998.

Whitehead, Alfred North. *Science and the Modern World*. New York: Free Press, 1967.

Wilson, Edward O. *Consilience: The Unity of Knowledge.* New York: Knopf, 1998.

Wolfram, Stephen. *A New Kind of Science.* Champaign, IL: Wolfram Media, 2002.

Wright, Robert. *Nonzero: The Logic of Human Destiny.* New York: Pantheon, 1999.

Photo Credits

Stellar "Fireworks Finale" Came First in the Young Universe
(Artist's Concept) Painting Credit: Adolf Schaller for STScI

Hubble Images Swarm of Ancient Stars
Image Credit: Hubble Heritage Team (AURA/ STScI/ NASA)

Hubble Reveals the Heart of the Whirlpool Galaxy
Image Credit: NASA and The Hubble Heritage Team (STScI/AURA)
Acknowledgment: N. Scoville (Caltech) and T. Rector (NOAO)

The Crab Nebula
Credit: NASA and The Hubble Heritage Team (STScI/AURA)
Acknowledgment: W. P. Blair (JHU)

Colorful Fireworks Finale Caps a Star's Life
Image Credit: NASA and The Hubble Heritage Team (STScI/AURA)
Acknowledgment: R. Fesen (Dartmouth) and J. Morse (Univ. of Colorado)

Multiple Generations of Stars in the Tarantula Nebula
Credit: Hubble Heritage Team (AURA / STScI / NASA)

Index

Index

Index

Index

Index

Index

About the Book

James N. Gardner presents a startling hypothesis for how our apparently bio-friendly universe began and what its ultimate destiny will be. Originally presented in peer-reviewed scientific journals, his radical "Selfish Biocosm" theory proposes that life and intelligence have not emerged in a series of random Darwinian accidents, but are essentially hardwired into the cycle of cosmic creation, evolution, death, and rebirth. He argues that the destiny of highly evolved intelligence (perhaps our distant progeny) is to infuse the entire universe with life, eventually to accomplish the ultimate feat of cosmic reproduction by spawning one or more "baby universes," which will themselves be endowed with life-generating properties.

In this explanation of the role of life in the cosmos, Gardner presents an eloquent and lucid synthesis of the most recent advances in physics, cosmology, biology, biochemistry, astronomy, and complexity theory. These disciplines increasingly find themselves approaching the frontier of what was once the exclusive province of philosophers and theologians. Gardner's Selfish Biocosm hypothesis challenges both Darwinists as well as advocates of intelligent design, and forces us to reconsider how we ourselves are shaping the future of life and the cosmos.

Inner Ocean Publishing

Expanding horizons
with books that
challenge the mind,
inspire the spirit,
and nourish the soul.

We invite you to visit us at:
www.innerocean.com

Inner Ocean Publishing, Inc.
PO Box 1239, Makawao
Maui, HI 96768, USA
Email: info@innerocean.com